Advances in Global Change Research

Volume 63

Editor-in-Chief
Martin Beniston, *University of Geneva, Switzerland*

Editorial Advisory Board
B. Allen-Diaz, *University of California, Berkeley, CA, USA*
W. Cramer, *Institut Méditerranéen de Biodiversité et d'Ecologie Marine et Continentale (IMBE), Aix-en-Provence, France*
S. Erkman, *Institute for Communication and Analysis of Science and Technology (ICAST), Geneva, Switzerland*
R. Garcia-Herrera, *Universidad Complutense, Madrid, Spain*
M. Lal, *Indian Institute of Technology, New Delhi, India*
U. Lutterbacher, *University of Geneva, Switzerland*
I. Noble, *Australian National University, Canberra, Australia*
M. Stoffel, *University of Bern, University of Geneva, Switzerland*
L. Tessier, *Institut Mediterranéen d'Ecologie et Paléoécologie (IMEP), Marseille, France*
F. Toth, *International Institute for Applied Systems Analysis (IIASA), Laxenburg, Austria*
M.M. Verstraete, *South African National Space Agency, Pretoria, South Africa*

More information about this series at http://www.springer.com/series/5588

Jessica E. Halofsky • David L. Peterson
Editors

Climate Change and Rocky Mountain Ecosystems

Editors
Jessica E. Halofsky
School of Environmental and Forest
  Sciences
University of Washington
Seattle, WA, USA

David L. Peterson
U.S. Forest Service
Pacific Northwest Research Station
Seattle, WA, USA

ISSN 1574-0919 ISSN 2215-1621 (electronic)
Advances in Global Change Research
ISBN 978-3-319-56927-7   ISBN 978-3-319-56928-4 (eBook)
DOI 10.1007/978-3-319-56928-4

Library of Congress Control Number: 2017945903

© Springer International Publishing AG 2018
This work is subject to copyright. All rights are reserved by the Publisher, whether the whole or part of the material is concerned, specifically the rights of translation, reprinting, reuse of illustrations, recitation, broadcasting, reproduction on microfilms or in any other physical way, and transmission or information storage and retrieval, electronic adaptation, computer software, or by similar or dissimilar methodology now known or hereafter developed.
The use of general descriptive names, registered names, trademarks, service marks, etc. in this publication does not imply, even in the absence of a specific statement, that such names are exempt from the relevant protective laws and regulations and therefore free for general use.
The publisher, the authors and the editors are safe to assume that the advice and information in this book are believed to be true and accurate at the date of publication. Neither the publisher nor the authors or the editors give a warranty, express or implied, with respect to the material contained herein or for any errors or omissions that may have been made. The publisher remains neutral with regard to jurisdictional claims in published maps and institutional affiliations.

Printed on acid-free paper

This Springer imprint is published by Springer Nature
The registered company is Springer International Publishing AG
The registered company address is: Gewerbestrasse 11, 6330 Cham, Switzerland

To Milo, Noah, Christina, and the next generation.

Foreword

Although earth scientists have been studying and discussing climatic variability for decades, climate change as a household phrase spread into public consciousness only recently. Awareness erupted in response to wake-up calls such as Al Gore's film, *An Inconvenient Truth*, produced in 2006, and the first United Nations Climate Change Conference in Copenhagen in 2009. Within land-managing agencies, such as the U.S. Forest Service and National Park Service, attention spread similarly. Hunger to learn about this emergent issue cycled a demand from resource managers on the ground back to scientists. The editors of this book, *Climate Change and Rocky Mountain Ecosystems*, as well as many of the chapter authors, responded by setting out on what turned into multiyear lecture circuits to field offices around the national forests, grasslands, and parks of the country.

What the scientists learned through those visits was as important as what they taught. They brought to the field audiences basic knowledge about weather and climate, interactions of climate and disturbances such as wildfire and insect epidemics, impacts of climate change on vegetation and wildlife, and the role of human actions in changing climates. What they heard resounded around one big question: What do we do now? In other words, how should land managers translate basic scientific information into relevant and practical actions on the ground? In those early years of discussion among scientists and managers about climate, the manual for addressing this fundamental question was unwritten and the toolkit empty.

Less than a decade later and the rich content of the current volume emerges, full of details on how to implement climate-smart resource management under the range of natural and institutional conditions encountered across landscapes of the Northern Rocky Mountains. Embracing aquatic to terrestrial ecosystems, plants to animals, and cultural resources to recreation, the 12 chapters elaborate strategies and tactics that connect the dots between science and practice in this vast ecoregion. Significantly, those early sessions of the lecture-circuit years set the stage for the underlying philosophy of effective climate adaptation promoted herein: the pivotal role of science-management partnerships. Then as now, teaching and learning reveal themselves as a multi-way process, with ideas flowing weblike among resource managers of different staff areas, among scientists of different disciplines, and among scientists and managers. Novel understanding, approaches, and tools emerged as a result of these interactions. If you want to know how to operationalize climate planning and practice in the Northern Rockies, read this book.

But wait! The stories and successes explained in this volume apply widely to other bioregions and institutional settings. The framework presented here, the lessons learned, and the library of climate-adaptation practices compiled are readily propagated elsewhere. The Northern Rockies Adaptation Partnership—the basic unit for experimentation and learning here—took an all-lands approach that spanned natural and social ecosystems from the cool-mesic western Rocky Mountains to the hot-dry rangelands and prairies of the eastern part of the region. With these partners came decades of collective experience for tackling and surmounting the many real challenges of resource management, as well as for innovating and implementing creative solutions. In the end, the reward for thinking and acting in climate-smart ways will be the heightened capacity of our wildlands, watersheds, and airsheds and those who live, play, and depend on them to effectively confront the climate challenges coming at them.

U.S. Forest Service
Albany, CA, USA

Connie Millar

Preface

Climate Change and Rocky Mountain Ecosystems describes the results of a cutting-edge effort to assess climate change vulnerabilities and develop adaptation options for ecosystems in the Northern Rocky Mountains region of the United States, focusing on national forests, grasslands, and parks in Northern Idaho, Montana, North Dakota, Northern South Dakota, and the Greater Yellowstone Ecosystem. Building on a framework developed in previous subregional climate change efforts, the Northern Rockies Adaptation Partnership (NRAP) was the first regional-scale, multi-resource climate change assessment in the United States. The NRAP was unprecedented in scale, scope, and breadth of the partnership, demonstrating the value of using a diverse science-management partnership and a consistent framework to assess climate change effects and identify on-the-ground adaptation options.

This book provides concise descriptions of state-of-science climate change vulnerability assessments for water, fisheries, vegetation, disturbance, wildlife, recreation, ecosystem services, and cultural resources in the Northern Rockies. Adaptation strategies and tactics, including both familiar and novel ecosystem management approaches, are described for all resource areas. Lessons learned and next steps are also described in a concluding chapter.

Chapter 1 provides an overview of ecosystems in the Northern Rockies region and outlines the NRAP vulnerability assessment and adaptation process. Chapter 2 describes historical climate and future climate projections for the Northern Rockies region and five subregions within. Chapters 3, 4, 5, 6, 7 and 8 provide detailed physical and ecological climate change vulnerability assessments for hydrology, fisheries, forest and rangeland vegetation, ecological disturbance, and wildlife. Chapters 9, 10 and 11 focus on climate change vulnerabilities for social values and resources including recreation, cultural heritage, and other ecosystem services. Far more than literature reviews, these assessments synthesize the best available science, evaluate the quality and relevance of the science for each application, and identify geographic locations where sensitivity is high. For several assessments, new climate impact model analyses were conducted. Related adaptation strategies and tactics are described in each chapter, except for disturbance and ecosystem services, for which adaptation options are integrated in other chapters. Finally,

Chap. 12 describes potential applications of the vulnerability assessment and opportunities for implementing adaptation options.

We are optimistic that the vulnerability assessments and adaptation options developed through the NRAP will result in revised management approaches on the ground. Follow-up projects are already developing in the region, and information on potential climate change effects and adaptation is being integrated in national forest plan revisions, which will help national forests comply with the U.S. Forest Service 2012 Planning Rule. These projects and applications demonstrate the value of enduring relationships built during the course of the NRAP that have increased the capability of federal agencies to incorporate climate change in resource management and planning.

Only 5 years ago, climate change readiness was barely visible in the western United States. Now, organizational capacity of federal land management is accelerating as a result of science-management partnerships such as the one described here. Addressing the effects of climate change on natural resources will be one of the great challenges for society in future decades. It is our hope that this book will help improve our understanding of how humans are affecting nature and motivate timely implementation of adaptation in the years ahead.

School of Environmental and Forest Sciences University of Washington Seattle, WA, USA	Jessica E. Halofsky
Pacific Northwest Research Station U.S. Forest Service Seattle, WA, USA	David L. Peterson

Acknowledgments

This effort would not have been possible without the energy and leadership of Linh Hoang. We thank our outstanding team of authors, who not only developed excellent scientific assessments but also took the time to participate in multiple workshops to help resource managers across the Northern Rockies understand the implications of climate change on resource management. We are grateful to land managers in the U.S. Forest Service and other agencies and organizations who participated in workshops and provided the adaptation options described here. We thank Linda Joyce and Joanne Ho for editorial assistance. We are also grateful to the USGS North Central Climate Science Center for providing state-of-science climate projections. Helpful reviews of previous versions of these chapters were provided by Polly Buotte, Michael Case, Sean Finn, Paul Gobster, Jeff Hicke, Darryll Johnson, Morris Johnson, Mark Muir, Fred Noack, Douglas Peterson, Christina Restaino, Patrick Saffel, Jose Sanchez, Douglas Shinneman, Nikola Smith, Scott Spaulding, Linda Spencer, Michael Sweet, Laurie Yung, and Paul Zambino. Financial support was provided by the Forest Service Office of Sustainability and Climate, Northern Region, Pacific Northwest Research Station, and Rocky Mountain Research Station, and by the National Oceanic and Atmospheric Administration. This book is a contribution of the Western Mountain Initiative.

Contents

1. **Assessing Climate Change Effects in the Northern Rockies** 1
   S. Karen Dante-Wood, David L. Peterson, and Jessica E. Halofsky

2. **Historical and Projected Climate in the Northern Rockies Region** 17
   Linda A. Joyce, Marian Talbert, Darrin Sharp, and John Stevenson

3. **Effects of Climate Change on Snowpack, Glaciers, and Water Resources in the Northern Rockies** 25
   Charles H. Luce

4. **Effects of Climate Change on Cold-Water Fish in the Northern Rockies** 37
   Michael K. Young, Daniel J. Isaak, Scott Spaulding, Cameron A. Thomas, Scott A. Barndt, Matthew C. Groce, Dona Horan, and David E. Nagel

5. **Effects of Climate Change on Forest Vegetation in the Northern Rockies** 59
   Robert E. Keane, Mary Frances Mahalovich, Barry L. Bollenbacher, Mary E. Manning, Rachel A. Loehman, Terrie B. Jain, Lisa M. Holsinger, and Andrew J. Larson

6. **Effects of Climate Change on Rangeland Vegetation in the Northern Rockies** 97
   Matt C. Reeves, Mary E. Manning, Jeff P. DiBenedetto, Kyle A. Palmquist, William K. Lauenroth, John B. Bradford, and Daniel R. Schlaepfer

7	**Effects of Climate Change on Ecological Disturbance in the Northern Rockies** 115
	Rachel A. Loehman, Barbara J. Bentz, Gregg A. DeNitto, Robert E. Keane, Mary E. Manning, Jacob P. Duncan, Joel M. Egan, Marcus B. Jackson, Sandra Kegley, I. Blakey Lockman, Dean E. Pearson, James A. Powell, Steve Shelly, Brytten E. Steed, and Paul J. Zambino
8	**Effects of Climate Change on Wildlife in the Northern Rockies** 143
	Kevin S. McKelvey and Polly C. Buotte
9	**Effects of Climate Change on Recreation in the Northern Rockies** 169
	Michael S. Hand and Megan Lawson
10	**Effects of Climate Change on Ecosystem Services in the Northern Rockies** 189
	Travis Warziniack, Megan Lawson, and S. Karen Dante-Wood
11	**Effects of Climate Change on Cultural Resources in the Northern Rockies** 209
	Carl M. Davis
12	**Toward Climate-Smart Resource Management in the Northern Rockies** 221
	Jessica E. Halofsky, David L. Peterson, S. Karen Dante-Wood, and Linh Hoang

Index 229

Contributors

Scott A. Barndt U.S. Forest Service, Gallatin National Forest, Bozeman, MT, USA

Barbara J. Bentz U.S. Forest Service, Rocky Mountain Research Station, Logan, UT, USA

Barry L. Bollenbacher U.S. Forest Service, Northern Region, Missoula, MT, USA

John B. Bradford U.S. Geological Survey, Southwest Biological Science Center, Flagstaff, AZ, USA

Polly C. Buotte College of Forestry, Oregon State University, Corvallis, OR, USA

S. Karen Dante-Wood U.S. Forest Service, Office of Sustainability and Climate, Washington, DC, USA

Carl M. Davis U.S. Forest Service, Northern Region, Missoula, Montana, USA

Gregg A. DeNitto U.S. Forest Service, Northern Region, Missoula, MT, USA

Jeff P. DiBenedetto U.S. Forest Service, Custer National Forest, Retired, Billings, MT, USA

Jacob P. Duncan Department of Mathematics and Statistics, Utah State University, Logan, UT, USA

Joel M. Egan U.S. Forest Service, Northern Region, Missoula, MT, USA

Matthew C. Groce U.S. Forest Service, Rocky Mountain Research Station, Boise, ID, USA

Jessica E. Halofsky School of Environmental and Forest Sciences, University of Washington, Seattle, WA, USA

Michael S. Hand U.S. Forest Service, Rocky Mountain Research Station, Washington, DC, USA

Linh Hoang U.S. Forest Service, Northern Region, Missoula, MT, USA

Lisa M. Holsinger U.S. Forest Service, Rocky Mountain Research Station, Missoula, MT, USA

Dona Horan U.S. Forest Service, Rocky Mountain Research Station, Boise, ID, USA

Daniel J. Isaak U.S. Forest Service, Rocky Mountain Research Station, Boise, ID, USA

Marcus B. Jackson U.S. Forest Service, Northern Region, Missoula, MT, USA

Terrie B. Jain U.S. Forest Service, Rocky Mountain Research Station, Moscow, ID, USA

Linda A. Joyce U.S. Forest Service, Rocky Mountain Research Station, Fort Collins, CO, USA

Robert E. Keane U.S. Forest Service, Rocky Mountain Research Station, Missoula, MT, USA

Sandra Kegley U.S. Forest Service, Northern Region, Missoula, MT, USA

Andrew J. Larson Department of Forest Management, University of Montana, Missoula, MT, USA

William K. Lauenroth Department of Botany, University of Wyoming, Laramie, WY, USA

Megan Lawson Headwaters Economics, Bozeman, MT, USA

I. Blakey Lockman U.S. Forest Service, Pacific Northwest Region, Portland, OR, USA

Rachel A. Loehman U.S. Geological Survey, Alaska Science Center, Anchorage, AK, USA

Charles H. Luce U.S. Forest Service, Rocky Mountain Research Station, Boise, ID, USA

Mary Frances Mahalovich U.S. Forest Service, Northern, Rocky Mountain, Southwestern, and Intermountain Regions, Moscow, ID, USA

Mary E. Manning U.S. Forest Service, Northern Region, Missoula, MT, USA

Kevin S. McKelvey U.S. Forest Service, Rocky Mountain Research Station, Missoula, CO, USA

David E. Nagel U.S. Forest Service, Rocky Mountain Research Station, Boise, ID, USA

Kyle A. Palmquist Department of Botany, University of Wyoming, Laramie, WY, USA

Dean E. Pearson U.S. Forest Service, Rocky Mountain Research Station, Missoula, MT, USA

David L. Peterson U.S. Forest Service, Pacific Northwest Research Station, Seattle, WA, USA

James A. Powell Department of Mathematics and Statistics, Utah State University, Logan, UT, USA

Matt C. Reeves U.S. Forest Service, Rocky Mountain Research Station, Missoula, MT, USA

Daniel R. Schlaepfer Department of Environmental Sciences, University of Basel, Basel, Switzerland

Darrin Sharp Oregon Climate Change Research Institute, Oregon State University, Corvallis, OR, USA

Steve Shelly U.S. Forest Service, Northern Region, Missoula, MT, USA

Scott Spaulding U.S. Forest Service, Northern Region, Missoula, MT, USA

Brytten E. Steed U.S. Forest Service, Northern Region, Missoula, MT, USA

John Stevenson Climate Impacts Research Consortium, Oregon State University, Corvallis, OR, USA

Marian Talbert Colorado State University, Fort Collins, CO, USA

Cameron A. Thomas U.S. Forest Service, Northern Region, Missoula, MT, USA

Travis Warziniack U.S. Forest Service, Rocky Mountain Research Station, Fort Collins, CO, USA

Meredith M. Webster U.S. Forest Service, Northern Region, Washington, DC, USA

Michael K. Young U.S. Forest Service, Rocky Mountain Research Station, Missoula, MT, USA

Paul J. Zambino U.S. Forest Service, Northern Region, Missoula, MT, USA

Chapter 1
Assessing Climate Change Effects in the Northern Rockies

S. Karen Dante-Wood, David L. Peterson, and Jessica E. Halofsky

Abstract The Northern Adaptation Partnership (NRAP) identified climate change issues relevant to resource management in the Northern Rockies (USA) region, and developed solutions that minimize negative effects of climate change and facilitate transition of diverse ecosystems to a warmer climate. The NRAP region covers 74 million hectares, spanning northern Idaho, Montana, northwest Wyoming, North Dakota, and northern South Dakota, and includes 15 national forests and 3 national parks across the U.S. Forest Service Northern Region and adjacent Greater Yellowstone Area. U.S. Forest Service scientists, resource managers, and stakeholders worked together over a two-year period to conduct a state-of-science climate change vulnerability assessment and develop adaptation options for national forests and national parks in the Northern Rockies region. The vulnerability assessment emphasized key resource areas—water, fisheries, wildlife, forest and rangeland vegetation and disturbance, recreation, cultural heritage, and ecosystem services—regarded as the most important for local ecosystems and communities. Resource managers used the assessment to develop a detailed list of ways to address climate change vulnerabilities through management actions. The large number of adaptation strategies and tactics, many of which are a component of current management practice, provide a pathway for slowing the rate of deleterious change in resource conditions.

Keywords Vulnerability assessment • Adaptation options • Northern Rockies • Climate change

---

S.K. Dante-Wood (✉)
U.S. Forest Service, Office of Sustainability and Climate, Washington, DC, USA
e-mail: skdante@fs.fed.us

D.L. Peterson
U.S. Forest Service, Pacific Northwest Research Station, Seattle, WA, USA
e-mail: peterson@fs.fed.us

J.E. Halofsky
School of Environmental and Forest Sciences, University of Washington, Seattle, WA, USA
e-mail: jhalo@uw.edu

© Springer International Publishing AG 2018
J.E. Halofsky, D.L. Peterson (eds.), *Climate Change and Rocky Mountain Ecosystems*, Advances in Global Change Research 63,
DOI 10.1007/978-3-319-56928-4_1

1.1 Introduction

The Northern Rocky Mountains—in this case, the portion within the United States—contain some of the most magnificent landscapes on Earth, stretching from high mountains to grasslands, from alpine glaciers to broad rivers (Fig. 1.1). Once inhabited solely by Native Americans, the region has been altered by two centuries of settlement by Euro-Americans, including extractive activities such as timber harvest, grazing, mining, and water diversions. A significant portion of the Northern Rockies is managed by federal agencies, including 15 national forests, 3 national parks, and the largest contiguous area of wilderness in the continental United States.

As "wild" as this region may seem, it is of course not immune to the effects of climate change. An increase in wildfire extent and large insect outbreaks, and their relationship to a warmer climate, have captured the attention of both natural resource managers and the general public. Federal agencies in the region have recognized that climate change will affect their ability to manage for the ecosystem services and values to which the public are accustomed. Federal leadership and resource managers in this region realize that timely adjustment of planning and management—through a "climate change lens"—will be needed to accomplish sustainable resource management in the future.

Recent focus on climate change in the Northern Rockies builds on prior assessment, and adaptation efforts in the western United States have demonstrated the value of science-management partnerships for increasing climate change awareness and facilitating adaptation on federal lands:

- Olympic National Forest and Olympic National Park (Washington) produced the first multi-resource assessment of climate change effects on federal lands, as well as adaptation options that are now being implemented (Halofsky et al. 2011; Littell et al. 2012).
- Tahoe National Forest, Inyo National Forest, and Devils Postpile National Monument held workshops and developed the Climate Project Screening Tool in order to incorporate adaptation into project planning (Morelli et al. 2012).
- Shoshone National Forest (Wyoming) synthesized information on past climate, future climate projections, and potential effects of climate change for multiple ecosystems (Rice et al. 2012).
- The North Cascadia Adaptation Partnership assessed resource vulnerabilities and developed adaptation options for two national forests and two national parks in Washington (Raymond et al. 2013, 2014).

In the largest effort to date in the eastern United States, Chequamegon-Nicolet National Forest (Wisconsin) conducted a vulnerability assessment for forest resources and developed adaptation options (Swanston et al. 2011; Swanston and Janowiak 2012). Finally, watershed vulnerability assessments, conducted on 11 national forests throughout the United States, were locally focused (at a national forest scale) and included water resource values, hydrologic response to climate change, watershed condition and landscape sensitivity (Furniss et al. 2013).

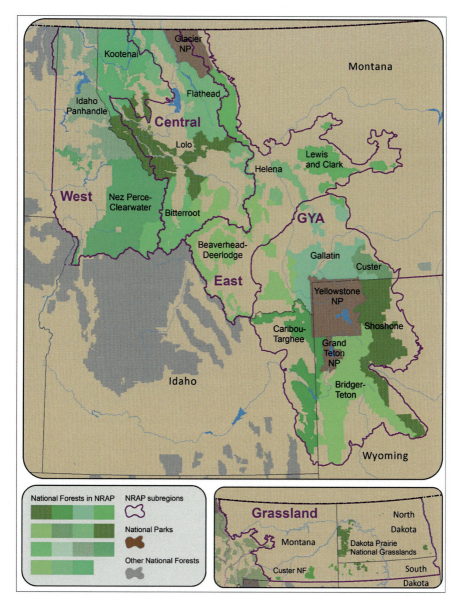

Fig. 1.1 National forests and national parks included in the climate change assessment for the Northern Rockies Adaptation Partnership (NRAP) (Map by R. Norheim)

A conceptual framework and process for conducting assessments and developing adaptation options on national forests have been well documented (Peterson et al. 2011; Swanston and Janowiak 2012). Five key steps guide this process:

1. Educate: Ensure that resource managers are aware of basic climate change science, integrating that understanding with knowledge of local conditions and issues.
2. Assess: Evaluate the sensitivity and adaptive capacity of natural and cultural resources to climate change.
3. Adapt: Develop management options for adapting resources and organizations to climate change.
4. Implement: Incorporate adaptation options and climate-smart thinking into planning and management.
5. Monitor: Evaluate the effectiveness of on-the-ground management and adjust as needed.

1.2 Northern Rockies Adaptation Partnership Process

The Northern Rockies Adaptation Partnership (NRAP) was created to address the potential effects of climate change in the context of ongoing ecosystem-based management and ecological restoration. Restoration is a priority in national forests, especially related to hazardous fuel reduction in dry forests (stand density reduction plus surface fuel removal), and restoration of riparian areas to improve hydrological and biological function. Restoration must be integrated with climate change assessment and adaptation to ensure long-term sustainability of ecosystems.

Initiated in 2013, the NRAP is a science-management partnership that includes U.S. Forest Service (USFS) regional offices and national forests; USFS Pacific Northwest and Rocky Mountain Research Stations; Glacier, Yellowstone, and Grand Teton National Parks; Great Northern and Plains and Prairie Potholes Landscape Conservation Cooperatives; Department of the Interior North Central Climate Science Center; Greater Yellowstone Coordinating Committee; Oregon State University; and EcoAdapt. By working collaboratively with scientists and resource managers and focusing on a specific region, the goal of NRAP was to provide the scientific foundation for operationalizing climate change in planning, ecological restoration, and project management in the Northern Rockies (Peterson et al. 2011; Swanston and Janowiak 2012; Raymond et al. 2013, 2014). Specific objectives were:

- Conduct a vulnerability assessment of the effects of climate change on hydrology, fisheries, wildlife, forested and non-forested vegetation and disturbance, recreation, cultural resources, and ecosystem services.
- Develop adaptation options that help reduce negative effects of climate change and assist the transition of biological systems and management to a changing climate.
- Develop an enduring science-management partnership to facilitate ongoing dialogue and activities related to climate change in the Northern Rockies.

Vulnerability assessments typically involve assessing exposure, sensitivity, and adaptive capacity (IPCC 2007), where exposure is the degree to which the system is exposed to changes in climate, sensitivity is an inherent quality of the system that indicates the degree to which it could be affected by climate change, and adaptive capacity is the ability of a system to respond and adjust to the exogenous influence of climate. Vulnerability assessments can be both qualitative and quantitative, focusing on whole systems or individual species or resources (Glick et al. 2011; Hansen et al. 2016). For the NRAP, we used scientific literature and expert knowledge to assess exposure, sensitivity, and adaptive capacity relative to key vulnerabilities in each resource area. The assessment process took place over 16 months, including monthly phone meetings for each of the resource-specific assessment teams. Each assessment team identified key questions, selected values to assess, and determined which climate change models best informed the assessment. In some cases, assessment teams conducted spatial analyses and/or ran and interpreted models, selected criteria in which to evaluate model outputs, and developed maps of model output and resource sensitivities.

After identifying key vulnerabilities for each resource sector, workshops were convened in October and November 2014 in Bismarck, North Dakota; Bozeman, Montana; Coeur d'Alene, Idaho; Helena, Montana; and Missoula, Montana to present and discuss the vulnerability assessment, and to elicit adaptation options from resource managers. During these workshops, scientists and resource specialists presented information on climate change effects and current management practices for each resource area. Information from the region-wide assessment was also downscaled to identify the most significant vulnerabilities to climate change for priority resources in each subregion. Facilitated dialogue was used to identify key sensitivities and adaptation options. Participants identified strategies (general approaches) and tactics (on-the-ground actions) for adapting resources and management practices to climate change, as well as opportunities and barriers for implementing these adaptation actions into projects, management plans, partnerships and policies. Participants focused on adaptation options that can be implemented given our current scientific understanding of climate change effects, but they also identified research and monitoring that would benefit future efforts to assess vulnerability and guide management. Facilitators captured information generated during the workshops with a set of spreadsheets adapted from Swanston and Janowiak (2012). Initial results from the workshops were augmented with continued dialogue with federal agency resource specialists. Detailed vulnerability assessment and adaptation results are described in a technical report (Halofsky et al. 2017).

1.3 Toward Implementation of Climate-Smart Management

The NRAP vulnerability assessment provides information on climate change effects needed for national forest and national park plans, project plans, conservation strategies, restoration, and environmental effects analysis. Climate change sensitivities and adaptation options developed at the regional scale provide the scientific foundation for subregional and national forest and national park vulnerability assessments, adaptation planning, and resource monitoring. We expect that over time, and as needs and funding align, appropriate adaptation options will be incorporated into plans and programs of federal management units. We also anticipate that resource specialists will apply this assessment in land management throughout the region, thus operationalizing climate-smart resource management and planning.

Adaptation planning is an ongoing and iterative process. Implementation may occur at critical times in the planning process, such as when managers revise USFS land management plans and other planning documents, or after the occurrence of extreme events and ecological disturbances (e.g., wildfire). We focus on adaptation options for the USFS and National Park Service (NPS), but this information can be used by other land management agencies as well. Furthermore, the approach used here can be emulated by agencies and organizations outside the Northern Rockies, thus propagating climate-smart management across larger areas.

The USFS and NPS climate change strategies identify the need to build partnerships and work across jurisdictional boundaries when planning for adaptation, that is, an "all-lands" approach. The NRAP is an inclusive partnership of multiple agencies and organizations with an interest in managing natural resources in a changing climate. In addition to representatives from the national forests, grasslands, and parks, several other agencies and organizations participated in the resource sector workshops. This type of partnership enables a coordinated and complementary approach to adaptation that crosses jurisdictional boundaries (Olliff and Hansen 2016). Communicating climate change information and engaging employees, partners, and the general public in productive discussions is also an integral part of successfully responding to climate change. Sharing climate change vulnerability assessments and adaptation strategies across administrative boundaries will contribute to the success of climate change responses throughout the Northern Rockies.

1.4 A Brief Tour of the Northern Rockies

The NRAP includes 15 national forests, 3.2 million hectares of wilderness, and 3 national parks across the USFS Northern Region and the adjacent Greater Yellowstone Area. The NRAP region covers 74 million hectares (Fig. 1.1), spanning northern Idaho, Montana, northwest Wyoming, North Dakota, and northern South Dakota. In order to capture the diversity of biogeography in this reagion, the NRAP climate change vulnerability assessment and adaptation strategy development process were conducted for five subregions:

- **Western Rockies:** Idaho Panhandle National Forest (NF), Kootenai NF, Nez Perce-Clearwater NF, Glacier National Park (NP)
- **Eastern Rockies:** Beaverhead-Deerlodge NF (eastern portion), Custer NF (eastern portion), Gallatin NF (northern portion), Helena NF, Lewis and Clark NF
- **Central Montana:** Bitterroot NF, Flathead NF, Lolo NF
- **Grassland:** Custer NF (part), Dakota Prairie Grasslands

- **Greater Yellowstone Area:** Bridger-Teton NF, Caribou-Targhee NF, Shoshone NF, Gallatin NF (southern portion), Custer NF (western portion), Beaverhead-Deerlodge NF (western portion), Grand Teton NP, Yellowstone NP

1.4.1 Western Rockies Subregion

The Western Rockies subregion, which occupies 7 million hectares, is extremely mountainous and heavily forested. It contains numerous large rivers, including the Salmon River which winds 680 km through central Idaho and provides habitat for Pacific salmon species. Other major rivers include the Clearwater, Kootenai, Pend Oreille, and Clark Fork of the Columbia (Fig. 1.2). Climate in this region is affected by a maritime atmospheric pattern; summers are hot and dry, and winters are relatively cold due to the high amount of moisture carried through the Columbia River Gorge.

Commercially harvested coniferous species in this area include Douglas-fir (*Pseudotsuga menziesii*), Engelmann spruce (*Picea engelmannii*), grand fir (*Abies grandis*), lodgepole pine (*Pinus contorta* var. *latifolia*), ponderosa pine (*P. ponderosa*), subalpine fir (*A. lasiocarpa*), western hemlock (*Tsuga heterophylla*), western larch (*Larix occidentalis*), western redcedar (*Thuja plicata*), and western white pine (*P. monticola*). Other species not used for wood products include whitebark pine (*P. albicaulis*), limber pine (*P. flexilis*), alpine larch (*Larix lyallii*), mountain hemlock (*Tsuga mertensiana*), and western juniper (*Juniperus occidentalis*). Quaking aspen (*Populus tremuloides*), black cottonwood (*P. nigra*) and paper birch (*Betula papyrifera*) are also commonly found. Common shrub species include serviceberry

Fig. 1.2 The Western Rockies subregion is characterized by complex mountainous topography with mixed conifer forests and streams (Photo by U.S. Forest Service)

(*Amelanchier alnifolia*), redosier dogwood (*Cornus sericea*), oceanspray (*Holodiscus discolor*), Lewis mockorange (*Philadelphus lewisii*), huckleberries (*Vaccinium* spp.) and smooth sumac (*Rhus glabra*) (Sullivan et al. 1986).

The Western Rockies provide habitat for over 300 animal species, including iconic mammals such as black bear (*Ursus americanus*), grizzly bear (*U. arctos*), elk (*Cervus elaphus*), moose (*Alces alces*), and gray wolf (*Canis lupus*). Avian taxa include bald eagle (*Haliaeetus leucocephalus*), golden eagle (*Aquila chrysaetos*), osprey (*Pandion haliaetus*), many species of owls, wild turkey (*Meleagris gallopavo*), California quail (*Callipepla californica*), and greater sage-grouse (*Centrocercus urophasianus*). Fish species include native cutthroat trout (*Oncorhynchus clarkii*), rainbow trout (*O. mykiss*), and bull trout (*Salvelinus confluentus*), and nonnative brook trout (*S. fontinalis*). The Kootenai River is home to the endangered white sturgeon (*Acipenser transmontanus*) and threatened burbot (*Lota lota*).

Wildfire is a dominant influence on the structure, function, and productivity of forest ecosystems in the Western Rockies, with stand replacement fires occurring at 50–500 year intervals, and surface fires occurring in dry forests at 2–50 year intervals. Frequent fires keep many forests in the early stages of succession as indicated by high numbers of western larch and pine (Schnepf and Davis 2013), although fire exclusion during the past 80 years has reduced fire frequency in lower-elevation dry forests, resulting in dense stands and elevated accumulation of surface fuels.

Mountain pine beetles (*Dendroctonus ponderosae*) kill large numbers of lodgepole pine, often in outbreaks of thousands of hectares, and they increasingly kill whitebark pine and limber pine (*P. flexilis*) at high elevation as the climate continues to warm. White pine blister rust (*Cronartium ribicola*), a nonnative fungus, causes mortality in five-needle pines (western white pine, whitebark pine, limber pine), and has greatly reduced the dominance of western white pine (Schwandt et al. 2013). Forests dominated by Douglas-fir and grand fir have increased as a result, accelerating forest succession toward shade tolerant, late-successional true firs, western hemlock, and western redcedar (Bollenbacher et al. 2014).

1.4.2 Central Rockies Subregion

The Central Rockies subregion, which occupies 5 million hectares, contains steep mountains, rolling meadows, large rivers, and lakes, and alpine ecosystems throughout its mountain ranges (Fig. 1.3). It also contains the largest contiguous area of designated wilderness in the United States outside of Alaska. The Bitterroot and Missoula Valleys located in west-central Montana experience an inland mountain climate. Air masses that develop over the Pacific Ocean release moisture in the Cascade Range and over the mountains of northern Idaho. West-central Montana occupies the rain-shadow area, receiving dried-out Pacific air and little moisture in the valley bottoms (Lackschewitz 1991). Climate in the Flathead and Glacier region is similar, influenced by the Pacific Maritime atmospheric pattern with warm, dry summers and wet, cold winters.

Fig. 1.3 The Central Rockies subregion is characterized by glacially carved topography, dense coniferous forest, and lakes in high mountain landscapes (Photo by U.S. Forest Service)

Microclimate has a big effect on the distribution and productivity of vegetation. Forests in the Bitterroot and Missoula valleys are drier than those in Idaho and northwestern Montana. Species found here include western redcedar, western white pine, Pacific yew (*Taxus brevifolia*), bride's bonnet (*Clintonia uniflora*), American trail plant (*Adenocaulon bicolor*), and threeleaf foamflower (*Tiarella trifoliata*). Intermountain forest species dominate the west-central Montana landscape, including western larch, alpine larch (*Larix lyallii*), ponderosa pine, mock azalea (*Menziesia ferruginea*), and common beargrass (*Xerophyllum tenax*). Bottomland ponderosa pine and hardwood species are found in moist sites, whereas different types of bunchgrass species (*Agropyron, Festuca*) and a mixture of ponderosa pine and bunchgrasses are found in dry sites. Douglas-fir, grand fir, and subalpine fir dominate at higher elevation (Lackschewitz 1991). In the Flathead Valley and Glacier National Park, lower elevations are dominated by Douglas-fir, ponderosa pine, grand fir, Engelmann spruce, and western redcedar. Douglas-fir, western larch and subalpine fir are common at mid elevation, and whitebark pine is found at high elevation (Newlon and Burns 2009). Black cottonwood and quaking aspen are common deciduous trees at lower elevations.

The Central Rockies contain over 60 species of mammals, with wilderness locations having relatively intact populations, including gray wolf, Canada lynx (*Lynx canadensis*), mountain lion (*Felis concolor*), mountain goat (*Oreamnos americanus*), bighorn sheep (*Ovis canadensis*), and grizzly bear. Hundreds of bird species are found in the Central Rockies, including killdeer (*Charadrius vociferus*) and spotted sandpiper (*Actitis macularius*) in riparian areas, song sparrows (*Melospiza melodia*) in grassland, and willow flycatcher (*Empidonax traillii*) and MacGillivray's warbler (*Geothlypis tolmiei*) in shrubby habitat. Rivers contain populations of native bull trout, westslope cutthroat trout (*Oncorhynchus clarki lewisi*), northern pike minnow (*Ptychocheilus oregonensis*), and largescale sucker (*Catostomus macrocheilus*).

Wildfires were fairly regular at lower and middle elevations in the Bitterroot and Missoula valleys prior to 1900. Seral western larch and lodgepole pine previously dominated north-facing slopes, but fire exclusion has led to increased dominance of shade tolerant species. As a result, silvicultural and prescribed burning treatments are being used to increase the distribution and abundance of seral tree and shrub species (Lackschewitz 1991). Wildfires are becoming increasingly expensive to control as they make their way around the forested landscape but also as they enter the wildland-urban interface. Mountain pine beetle has caused serious damage and mortality to whitebark pine over the past century (Bartos and Gibson 1990). White pine blister rust has caused extensive mortality in whitebark pine, especially in Glacier National Park and adjacent areas, where over 70% of the trees are infected and 30% have died.

1.4.3 Eastern Rockies Subregion

The Eastern Rockies subregion (Fig. 1.1), in central and southwest Montana, contains coniferous forests on the eastern side of the Continental Divide, extending from high mountains in the west to broad plains in the east, including several large wilderness areas. Climate varies, based on location relative to the Continental Divide. The western side receives more precipitation as air masses from the west cool and release moisture over the mountain ranges, whereas on the eastern side, the air becomes warmer and drier, often accompanied by downslope air movement (Chinook winds) (Phillips 1999); the eastern portion of the subregion experiences a drier continental climate.

Numerous rivers flow through the Eastern Rockies, including the Missouri, Blackfoot, and Smith Rivers. The longest river in North America, the Missouri, begins at the confluence of the Jefferson and Madison Rivers near Three Forks, Montana and includes three reservoirs (Canyon Ferry, Hauser, and Upper Holter). These rivers are known for their blue-ribbon trout fishery status, scenic floats, and other water-based recreational activities.

Vegetation varies as a function of elevation and aspect. Lower elevations are dominated by grassland and sagebrush steppe that include needle-and-thread grass (*Hesperostipa comata*), Idaho fescue (*Festuca idahoensis*), sagebrush (*Artemisia* spp.), rabbitbrush (*Ericameria nauseosa*), and many herbaceous species. Fremont's cottonwood (*Populus fremontii*), yellow willow (*Salix lutea*), coyote willow (*S. exigua*), woods rose (*Rosa woodsii*), and golden currant (*Ribes aureum*) are found along rivers and streams. Dominant species in foothills and woodlands include limber pine, Rocky Mountain juniper (*Juniperus scopulorum*), Douglas-fir, and ponderosa pine. Understory species include antelope bitterbrush (*Purshia tridentata*), mountain-mahogany (*Cercocarpus* spp.), and skunkbrush sumac (*Rhus aromatica* var. *simplicifolia*). Douglas-fir and ponderosa pine dominate upper montane slopes, and lodgepole pine and Engelmann spruce are common at high elevation (Phillips 1999).

Iconic mammal species include mountain goat, bighorn sheep, elk, mountain lion, Canada lynx, wolverine (*Gulo gulo*), and black bear. Bird species include bald eagle, greater sage-grouse, peregrine falcon (*Falco peregrinus*), and red-tailed hawk (*Buteo jamaicensis*). Fly fishing opportunities in rivers and streams are plentiful due to abundant populations of westslope cutthroat trout, rainbow trout, brook trout, and northern pike (*Esox lucius*).

Forests have been subject to widespread drought, wildfire, and insect outbreaks over the past 20 years (Montana DEQ 2013). Several large wildfires have burned with uncharacteristic intensity because the absence of fire for several decades has resulted in elevated fuel accumulation. Mountain pine beetle has caused extensive mortality of lodgepole pine and some ponderosa pine, resulting from the concurrence of older, non-vigorous stands and elevated beetle populations caused by warmer temperature. Western spruce budworm has caused mortality and stunted growth in Engelmann spruce and Douglas-fir in some areas.

1.4.4 Greater Yellowstone Area Subregion

The Greater Yellowstone Area (GYA) subregion (Fig. 1.1), which occupies 9 million hectares, is defined by a group of 24 conterminous mountain ranges that wrap around the Yellowstone Plateau (Morgan 2007). The Yellowstone "hotspot" of thermal activity and associated geological forces have shaped the geography, topography, climate, soils, and biota of the GYA. The heat of eruptions that created calderas in this area provides the source for hot springs and geysers, one of the primary reasons that Yellowstone National Park was established; more geysers are found here than anywhere else in the world (NPS 2015).

The GYA is the source of the Missouri/Mississippi, Snake/Columbia, and Green/Colorado River systems. The Missouri River begins in the northwest corner of the GYA and merges into the Mississippi River, the Snake River begins in the southeast corner of the GYA and merges into the Columbia River, and the Green River is the main tributary of the Colorado River. Climate is predominantly continental in this subregion, with mild summers accompanied by thunderstorms, and cold winters with heavy snow at high elevations.

The GYA is one of the largest relatively intact and functional natural ecosystems in the temperate zone (Keiter and Boyce 1991). Valley bottoms are generally occupied by lodgepole pine (NPS 2013), lower slopes and richer soils support Engelmann spruce, subalpine fir, and Douglas-fir (Morgan 2007), and the highest elevations are dominated by whitebark pine (NPS 2012). Quaking aspen, willows, and cottonwoods are also found on valley toe slopes and bottoms. Mountain big sagebrush (*Artemisia tridentata* var. *vaseyana*) and Idaho fescue dominate lower elevation grassland and meadows (NPS 2012).

The GYA appears to have retained most of its historical complement of vertebrate wildlife species (NPS 2013), including significant elk and bison (*Bison bison*) herds. This ecosystem supports other megafauna such as grizzly bear, moose, white-tailed deer (*Odocoileus virginianus*), gray wolf, and coyote (*Canis latrans*). Distinctive avifauna include trumpeter swan (*Cygnus columbianus*) and bald eagle. Several hundred species of other small mammals and birds, and thousands of species of insects and other invertebrates occupy the subregion (Keiter and Boyce 1991).

Of the many fish species found in the GYA, westslope cutthroat trout and Yellowstone cutthroat trout (*Oncorhynchus clarkii bouvieri*) are keystone species preyed upon by many wildlife species. Cutthroat trout are at risk from hybridization and competition with nonnative lake trout (*Salvelinus namaycush*) (NPS 2012).

Wildfire has had an enormous impact in the subregion, especially the fires of 1988, which burned 500,000 hectares. These fires helped change how scientists, resource managers, and the general public think about the role of large fires in the fire ecology of Western forests. Mountain pine beetles have killed thousands of hectares of lodgepole pine in the GYA, although the outbreaks have not been as extensive as in other areas of North America.

1.4.5 Grassland Subregion

The Grassland subregion (Fig. 1.1), which occupies 48 million hectares (mostly on private land), extends across portions of Montana, Wyoming, North Dakota and South Dakota, where the rolling topography was influenced by glacial activity. The Missouri, Red, and Souris (or Mouse) Rivers are the three major river systems, in addition to the Tongue and Powder Rivers that the flow into the Yellowstone River in the southern portion of the subregion.

Three very different ecosystem types exist here: badlands, prairie, and ponderosa pine forest. The pine forests are found in "islands of green in a sea of rolling prairie." Most of the ponderosa pine forest in the Custer National Forest burned in the last decade. Located mostly along the Little Missouri River, the badlands are a collection of rugged landscapes as described above. Sioux Indians called the badlands "makosika" (land bad), and the French explorers called it "les mauvais terrers a traverser" which means bad lands to travel across (Bluemle 1996).

Vegetation in the badlands is dominated by grasses, forbs, shrubs, and small trees. Shortgrasses are dominant because annual precipitation is less than 30 cm. Common bunchgrasses include little bluestem *(Schizachyrium scoparium)*, blue grama *(Bouteloua gracilis)*, and needle-and-thread *(Hesperostipa comata)*. Open areas that can retain moisture throughout the year can support deciduous trees, such as eastern cottonwood *(Populus deltoides)*, green ash *(Fraxinus pennsylvanica)*, boxelder *(Acer negundo)* and American elm *(Ulmus americana)*. Drainages dominated by trees provide habitat for shrubs such as western serviceberry and chokecherry *(Prunus virginiana)*. Shrubs found in drier locations include sagebrush, winterfat *(Krascheninnikovia lanata)*, rabbitbrush, and buckbrush *(Ceanothus cuneatus)*. The badlands landscape provides wildlife habitat in native prairie, sagebrush, woody draws, shrubby areas, and buttes. This habitat supports several large

1 Assessing Climate Change Effects in the Northern Rockies

Fig. 1.4 The Grassland subregion is characterized by both native prairie (*left*) and badlands (*right*) (Photos by U.S. Forest Service)

mammals (bighorn sheep, pronghorn, elk, white-tailed deer, and mule deer) and a broad range of birds, reptiles, and amphibians.

The native prairie component of the Grassland subregion is dominated by grasses and forbs that tolerate low precipitation, strong winds, cold winters and hot summers, frequent wildfire, and herbivory (Fig. 1.4). Native grasses, which have extensive root systems that allow them to persist under stressful conditions (Herman and Johnson 2008a) include tallgrass, mixed grass, and shortgrass prairie. Most prairies have been largely replaced by agriculture because of their fertile soils. American bison was formerly the dominant herbivore on the prairies, and Native Americans depended on bison for their livelihood. However, Euro-American settlement and hunting nearly drove the bison to extinction. Other mammals found in prairie habitat include elk, red fox (*Vulpes vulpes*), coyote, and many smaller species. The Grassland subregion is home to over 100 species of fish, including northern pike, walleye (*Sander vitreus*), and sauger (*Sander canadensis*). Walleye inhabit large reservoirs such as Lake Sakakawea and Lake Oahe, while sauger are mostly found in the Missouri River (Herman and Johnson 2008b).

References

Bartos, D. L., & Gibson, K. E. (1990). Insects of whitebark pine with emphasis on mountain pine beetle. In: W. C. Schmidt, & K. J. McDonald (Comps.), *Proceedings, symposium on whitebark pine ecosystems: Ecology and management of a high-mountain resource* (General Technical Report INT-GTR-270, pp. 171–179). Ogden: U.S. Forest Service, Intermountain Research Station.

Bluemle, J. P. (1996). North Dakota's badlands. *Geology Today, 12*, 217–223.

Bollenbacher, B. L., Graham, R. T., & Reynolds, K. M. (2014). Regional forest landscape restoration priorities: Integrating historical conditions and an uncertain future in the Northern Rocky Mountains. *Journal of Forestry, 112*, 474–483.

Furniss, M. J., Roby, K. B., Cenderelli, D., et al. (2013). *Assessing the vulnerability of watersheds to climate change: Results of national forest watershed vulnerability pilot assessments* (General Technical Report PNW-GTR-884). Portland: U.S. Forest Service, Pacific Northwest Research Station.

Glick, P., Stein, B. A., & Edelson, N. A. (Eds.). (2011). *Scanning the conservation horizon: a guide to climate change vulnerability assessment*. Washington, DC: National Wildlife Federation.

Halofsky, J. E., Peterson, D. L., O'Halloran, K. A., & Hoffman, C. H. (2011). *Adapting to climate change at Olympic National Forest and Olympic National Park* (General Technical Report PNW-GTR-844). Portland: U.S. Forest Service, Pacific Northwest Research Station.

Halofsky, J. E., Peterson, D. L., Dante-Wood, S. K., et al. (Eds.). (2017). *Climate change vulnerability and adaptation in the Northern Rocky Mountains* (General Technical Report RMRS-XXX). Fort Collins: U.S. Forest Service, Rocky Mountain Research Station.

Hansen, A. J., Monahan, W. B., Theobald, D. M., & Olliff, S. T. (Eds.). (2016). *Climate change in wildlands*. Washington, DC: Island Press.

Herman, G. S., & Johnson, L. A. (2008a). *Prairie*. Bismarck: North Dakota Center for Distance Education/State Game and Fish Department.

Herman, G. S., & Johnson, L. A. (2008b). *Riparian areas*. Bismarck: North Dakota Center for Distance Education/State Game and Fish Department.

Intergovernmental Panel on Climate Change (IPCC). (2007). *Climate change 2007—The physical science basis. Climate Change 2007 working group 1 contribution to the fourth assessment report of the IPCC*. New York: Cambridge University Press.

Keiter, R. B., & Boyce, M. S. (1991). *The Greater Yellowstone ecosystem: Redefining America's wilderness heritage*. New Haven: Yale University Press.

Lackschewitz, K. (1991). *Vascular plants of west-central Montana—identification guidebook* (General Technical Report INT–277). Ogden: U.S. Forest Service, Intermountain Research Station.

Littell, J. S., Peterson, D. L., Millar, C. I., & O'Halloran, K. A. (2012). U.S. national forests adapt to climate change through science-management partnerships. *Climatic Change, 110*, 269–296.

Montana Department of Environmental Quality (DEQ). (2013). *Climate change and forestry: Montana's forests in an era of climate change*. Helena: Montana Department of Environmental Quality.

Morelli, T. L., Yeh, S., & Smith, N. M. (2012). *Climate project screening tool: An aid for climate change adaptation* (Research Paper PSW-RP-263). Albany: U.S. Forest Service, Pacific Southwest Research Station.

Morgan, L. A., (Ed.). (2007). *Integrated geoscience studies in the greater Yellowstone area—volcanic, tectonic, and hydrothermal processes in the Yellowstone geoecosystem* (Professional Paper 1717). Washington, DC: U.S. Geological Survey.

National Park Service (NPS). (2012). *Revisiting Leopold: Resource stewardship in the National Parks*. National Park System Advisory Board Science Committee. http://www.nps.gov/calltoaction/PDF/LeopoldReport_2012.pdf. 21 July 2015.

National Park Service (NPS). (2013). *Yellowstone resources and issues handbook: Greater Yellowstone Ecosystem*. Yellowstone National Park: National Park Service, Yellowstone National Park. http://www.nps.gov/yell/planyourvisit/upload/RI_2014_05_GYE.pdf. 8 Jan 2015.

National Park Service (NPS). (2015). *Geysers and how they work*. Yellowstone National Park: U.S. National Park Service, Yellowstone National Park. http://www.nps.gov/yell/naturescience/geysers.htm. 8 Jan 2015.

Newlon, K., & Burns, M. D. (2009). *Wetlands of the Flathead Valley: Change and ecological functions* (Report to the Montana Department of Environmental Quality and U.S. Environmental Protection Agency). Helena: Montana Natural Heritage Program.

Olliff, S. T., & Hansen, A. J. (2016). Challenges and approaches for integrating science into federal land management. In A. J. Hansen, W. B. Monahan, D. M. Theobald, & S. T. Olliff (Eds.), *Climate change in wildlands* (pp. 33–52). Washington, DC: Island Press.

Peterson, D. L., Millar, C.I., Joyce, L.A., et al. (2011). *Responding to climate change in national forests: A guidebook for developing adaptation options* (General Technical Report PNW-GTR-855). Portland: U.S. Forest Service, Pacific Northwest Research Station.

Phillips, W. (1999). *Central Rocky Mountain wildflowers: Including Yellowstone and Grand Teton National Parks*. Guilford: Globe Pequot Press.

Raymond, C. L., Peterson, D. L., & Rochefort, R. M. (2013). The North Cascadia Adaptation Partnership: A science-management collaboration for responding to climate change. *Sustainability, 5*, 136–159.

Raymond, C. L., Peterson, D. L., & Rochefort, R. M. (Eds.). (2014). *Climate change vulnerability and adaptation in the North Cascades region, Washington* (General Technical Report PNW-GTR-892). Portland: U.S. Forest Service, Pacific Northwest Research Station.

Rice, J., Tredennick, A., & Joyce, L. A. (2012). *Climate change on the Shoshone National Forest, Wyoming: A synthesis of past climate, climate projections, and ecosystem implications* (General Technical Report RMRS-GTR-274). Fort Collins: U.S. Forest Service, Rocky Mountain Research Station.

Schnepf, C., & Davis, A. S. (2013). Tree planting in Idaho. *Tree Planters Notes, 56*, 19–26.

Schwandt, J., Kearns, H., & Byler, J. (2013). *White pine blister rust–general ecology and management* (Insect and Disease Management Series 14.2). Washington: U.S. Forest Service, Forest Health Protection and State Forestry Organizations.

Sullivan, J., Long, L. E., & Menser, H. A. (1986). *Native plants from northern Idaho* (Bulletin 657). Moscow: University of Idaho, College of Agriculture.

Swanston, C. W., & Janowiak, M. K. (Eds.). (2012). *Forest adaptation resources: Climate change tools and approaches for land managers* (General Technical Report NRS-GTR-87). Newtown Square: U.S. Forest Service, Northern Research Station.

Swanston, C., Janowiak, M., Iverson, L., et al. (2011). *Ecosystem vulnerability assessment and synthesis: A report from the climate change response framework project in northern Wisconsin* (General Technical Report NRS-82). Newtown Square: U.S. Forest Service, Northern Research Station.

Chapter 2
Historical and Projected Climate in the Northern Rockies Region

Linda A. Joyce, Marian Talbert, Darrin Sharp, and John Stevenson

Abstract Climate influences the ecosystem services we obtain from forest and rangelands. An understanding of how climate may change in the future is needed to consider climate change in resource planning and management. In this chapter, we present the current understanding of the future changes in climate for the Northern Rockies region. Projected climate was derived from climate models in the Coupled Model Intercomparison Project version 5 (CMIP5) database, which was used in the most recent Intergovernmental Panel on Climate Change reports. Climate models project that the Earth's current warming trend will continue throughout the twenty-first century in the Northern Rockies. Compared to observed historical temperature, average warming across the Northern Rockies is projected to be about 2–3 °C by 2050, depending on greenhouse gas emissions. Seasonally, projected winter maximum temperature begins to rise above freezing in the mid-twenty-first century in several parts of the region. Projections for precipitation suggest a slight increase in the future, but precipitation projections, in general, have much higher uncertainty than those for temperature.

Keywords Precipitation • Temperature • Representative concentration pathways • CMIP5

L.A. Joyce (✉)
U.S. Forest Service, Rocky Mountain Research Station, Fort Collins, CO, USA
e-mail: ljoyce@fs.fed.us

M. Talbert
Colorado State University, Fort Collins, CO, USA
e-mail: Marian.K.Talbert@aphis.usda.gov

D. Sharp
Oregon Climate Change Research Institute, Oregon State University, Corvallis, OR, USA
e-mail: dsharp@coas.oregonstate.edu

J. Stevenson
Climate Impacts Research Consortium, Oregon State University, Corvallis, OR, USA
e-mail: jstevenson@coas.oregonstate.edu

© Springer International Publishing AG 2018
J.E. Halofsky, D.L. Peterson (eds.), *Climate Change and Rocky Mountain Ecosystems*, Advances in Global Change Research 63,
DOI 10.1007/978-3-319-56928-4_2

2.1 Introduction

Climate influences the ecosystem services that our society obtains from forest and rangeland ecosystems. Climate is described by the long-term characteristics of precipitation, temperature, wind, snowfall, and other measures of weather in a particular place. Day to day implementation of resource management practices are made in response to weather conditions; resource management strategies and plans are developed using our understanding of climate, the long-term average conditions. With the need to consider climate change in planning and management, an understanding of how climate may change in the future in a resource management planning area is valuable. In this chapter, we present the current understanding of potential future changes in climate for the Northern Rockies region.

Climate within the Northern Rockies region is influenced by the warm, wet maritime airflows from the Pacific Ocean and the cooler, drier airflows from Canada. In the Western, Central, Eastern, and Greater Yellowstone Area subregions (see Fig. 1.1 in Chap. 1), climate, especially at local scales, is strongly influenced by interactions among topography, elevation, and aspect. On the eastern edge of the Northern Rockies region, the Grassland subregion is influenced primarily by the cooler, drier airflows from Canada. Consequently, there are broad east-west changes in precipitation seasonality and amount, as well as strong elevation influences on temperatures. Trends and drivers for climatic variations will differ greatly from east to west.

2.2 Climate Model Overview

Global climate models have been used to understand the nature of global climate, how the atmosphere interacts with the ocean and the land surface. Scientists can use these models to pose questions about how changes in atmospheric chemistry affect global temperature and precipitation patterns. Given a set of plausible greenhouse gas emission scenarios, these models can be used to project potential future climate. These projections can be helpful in understanding how the environmental conditions of plants and animals might change in the future; how streamflow might vary with precipitation and timing of snowmelt; how wildfire, insects and disease outbreaks might be affected by changes in climate; and how humans might respond in their use of the outdoors and natural resources.

The Coupled Model Intercomparison Project (CMIP) began in 1995 to coordinate a common set of experiments for evaluating changes to past and future global climate (Meehl et al. 2007). This approach allows comparison of results from different global climate models around the world and improves our understanding of the "range" of possible climate change. The third CMIP modeling experiments, or CMIP3, were used in the Intergovernmental Panel on Climate Change (IPCC) Fourth Assessment Report (IPCC 2007), whereas CMIP5, the latest experiments, were used in the IPCC Fifth Assessment Report (IPCC 2013). The CMIP3 simulations of the twenty-first century were forced with emission scenarios from the Special Report on Emissions Scenarios (SRES) (Nakićenović et al. 2000). The CMIP5 simulations of the twenty-first century are driven by representative concentration pathways (RCPs) (van Vuuren et al. 2011). The RCPs do not define emissions, but instead define concentrations of greenhouse gases and other agents influencing the climate system. RCPs represent the range of current estimates regarding the evolution of radiative forcing, the total amount of extra energy entering the climate system throughout the twenty-first century and beyond. Projections made with RCP 2.6 show a total radiative forcing increase of 2.6 Wm^{-2} by 2100; projected increased radiative forcing through the scenarios of RCP 4.5, RCP 6.0 and RCP 8.5 indicate increases of 4.5, 6.0, and 8.5 Wm^{-2}, respectively. Unlike the SRES scenarios used in CMIP3, the RCPs in CMIP5 do not assume any particular climate policy actions.

2.3 Methods Used to Assess Future Climate in the Northern Rockies Region

In this chapter, we use results from the CMIP5 climate models to explore potential changes in the climate of the Northern Rockies region. Because output from global climate models is generally too coarse to represent climate dynamics in subregions and management areas relevant for the Northern Rockies, we utilized one of the many methods to bring climate projection information down to a scale that can be helpful to resource managers. We drew on climate projections that had been downscaled using the bias-correction and spatial disaggregation (BCSD) method (Maurer et al. 2007). Historical modeled and projected monthly temperature and precipitation for the 1950–2099 period were obtained from the Climate and Hydrology Projections archive at http://gdo-dcp.ucllnl.org/downscaled_cmip_projections. We used projections from 36 climate models for RCP 4.5 and 34 climate models for RCP 8.5 (Joyce et al. 2017). Spatial resolution of the data is 1/8-degree latitude-longitude and covers the entire Northern Rockies region.

Many of the resource chapters in this book drew on the CMIP3 projections that have been widely used in other assessments, such as the National Climate Assessment (Walsh et al. 2014), and the Forest Service Resource Planning Act Assessment (U.S. Department of Agriculture Forest Service 2012). Climate projections by Littell et al. (2011) have been used widely in the Pacific Northwest, hence we compared the CMIP5 results with the CMIP3 projections of Littell et al. (2011). For the Northern Rockies Region, projected change in temperature by the 2040–2060 period ranges from just under 1.1 °C to nearly 4.4 °C, with greater projected change under the RCP 8.5 scenario than the RCP 4.5 scenario. Change in precipitation across these CMIP5 models ranges from a decrease of 5% to an increase of 25% with a mean projected change of around 6% and 8% for RCP 4.5 and RCP 8.5, respectively. We conclude that the CMIP3 results for this region are in the same temperature range for the 2040–2060 period as the CMIP5 results presented here, but the CMIP5 precipitation projections are slightly wetter in the future (Joyce et al. 2017).

To report on the CMIP5 results for the Northern Rockies region, we used a base period of 1970–2009 for the historical climate, and compare projections for two periods (2030–2059, 2070–2099) with this historical climate (Fig. 2.1). These time periods were selected in an attempt to summarize climate that has influenced the current conditions (base period) and two future periods that will be relevant to long-term management action (such as road construction, hydrological infrastructure [see Chap. 3], or vegetation planting [see Chap. 5]). We report on the potential variability in projected climate across the Northern Rockies region by summarizing temperature and precipitation in the five subregions: Western Rockies, Central Rockies, Eastern Rockies, Greater Yellowstone Area, and Grassland (see Fig. 1.1 in Chap. 1). Data analysis was carried out in R (R Core Team 2016).

2.4 Projected Future Climate in the Northern Rockies

All subregions in the Northern Rockies will see increasingly warmer temperatures through the twenty-first century (Fig. 2.1). The historical map reflects the cooler temperatures in the mountainous regions, with the Greater Yellowstone Area subregion the coolest area and the Grassland subregion the warmest (Fig. 2.1). All areas are projected to warm under both RCPs, but warming is greater under RCP 8.5. Projections for precipitation suggest a slight increase in the future. However, precipitation projections, in general, have much higher uncertainty than those for temperature.

In the Western Rockies subregion, mean temperature is projected to increase 2.8–5.6 °C by 2100. Historically, winter, spring, and autumn minimum temperatures have been below freezing, a biologically important threshold. Spring minimum temperatures rise above freezing by mid-twenty-first century for RCP 8.5 and by 2080 in the RCP 4.5 scenario. Winter minimum temperatures remain below freezing in both future scenarios. However, maximum temperatures for winter, historically just below freezing, rise above freezing in both scenarios by the end of the century. Seasonal precipitation is projected to be slightly wetter in winter and spring, and slightly drier in summer.

In the Central Rockies subregion, annual mean monthly minimum temperature is projected to increase 3.3–6.7 °C, and annual mean monthly maximum temperature is projected to increase 2.8–6.1 °C by 2100. Summer mean maximum temperatures are projected to rise 2.8–6.5 °C, with the projected temperatures for the RCP 8.5 scenario outside of the historical ranges. Mean monthly minimum temperature (spring and autumn) and the mean monthly maximum temperature (winter), all historically below freezing, may rise above freezing by mid- to late-century. Seasonal precipitation is projected to be slightly wetter in winter and spring, and slightly drier in summer.

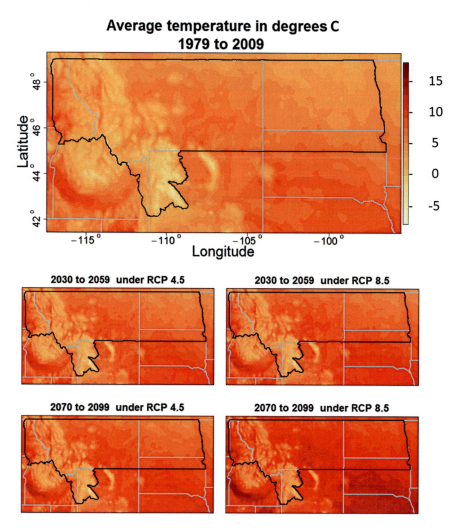

Fig. 2.1 Historical (1970–2009) and projected (2030–2059 and 2070–2099) mean annual monthly temperature for the Northern Rockies region under RCP 4.5 and RCP 8.5 scenarios. Projected climate results are the mean of 36 models for RCP 4.5 and 34 models for RCP 8.5. Spatial resolution of the data is 1/8-degree latitude-longitude

By 2100, annual mean monthly minimum temperature in the Eastern Rockies subregion is projected to increase 3.3–6.1 °C, and annual mean monthly maximum temperature is projected to increase 2.8–6.1 °C. Mean monthly maximum and minimum temperatures are projected to increase for all seasons. Mean monthly minimum temperature (spring and autumn) have historically been below freezing; these seasonal temperatures are projected to increase 2.8 °C for RCP 4.5, resulting in temperatures around freezing by end of twenty-first century. For the warmer scenario, summer maximum temperatures are projected to increase by 5.5 °C. The majority of the model projections rise above the historical range by the end of the century.

In the Greater Yellowstone Area subregion, annual mean monthly minimum temperature is projected to increase 2.8–5.6 °C, and annual mean monthly maximum temperature is projected to increase 3.9–6.7 °C by 2100. Winter maximum temperature is projected to rise above freezing in the mid-twenty-first century. Projected summer temperature is projected to increase 2.8 °C by 2060 and 5.6 °C by 2100. The Greater Yellowstone Area subregion is an area where changes at the local scale may differ from these broader estimates because of the complexities of topography, elevation, and aspect. These terrain complexities as well as snowpack conditions may provide areas of refugia for both plants and animals as climate changes.

For the Grassland subregion, warming trends indicate that future climate will be similar to the area south of this region. There is a pattern of a drier west and wetter east, with the average of climate models showing a slight shift to more of the wetter east. However, even with little or no change in precipitation, there is the potential for summer drying or drought caused by increased heat and increased evapotranspiration. Summer maximum temperatures increase by more than 6.5 °C; the majority of the projections by 2100 are outside of the historical range of maximum summer temperatures. Early snowmelt from the west will imply changes in streamflow and temperature, and therefore reservoir management and stream ecology.

Acknowledgements We acknowledge the World Climate Research Programme's Working Group on Coupled Modelling, which is responsible for CMIP, and we thank the climate modeling groups for producing and making available their model output. For CMIP the U.S. Department of Energy's Program for Climate Model Diagnosis and Intercomparison provides coordinating support and led development of software infrastructure in partnership with the Global Organization for Earth System Science Portals. Support for this chapter came from the Department of Interior's North Central and Northwest Climate Science Centers and the U.S. Forest Service Rocky Mountain Research Station. We thank Jeff Morisette for his participation in the climate analysis (Joyce et al. 2017). Any use of trade, product, or firm names is for descriptive purposes only and does not imply endorsement by the U.S. Government.

References

Intergovernmental Panel on Climate Change (IPCC). (2007). *Climate change 2007—The physical science basis. Climate change 2007 working group I contribution to the fourth assessment report of the IPCC*. New York: Cambridge University Press.

Intergovernmental Panel on Climate Change (IPCC). (2013). *Climate change 2013—The physical science basis. Contribution of working group I to the fifth assessment report of the intergovernmental panel on climate change*. Cambridge/New York: Cambridge University Press.

Joyce, L. A., Talbert, M., Sharp, D., et al. (2017). Historical and projected climate in the Northern Rockies Adaptation Partnership Region. In: J. E. Halofsky, D. L. Peterson, S. K. Dante-Wood, et al. (Eds.), *Climate change vulnerability and adaptation in the Northern Rocky Mountains* (General Technical Report RMRS-XXX). Fort Collins: U.S. Forest Service, Rocky Mountain Research Station.

Littell, J. S., Elsner, M. M., Mauger, G., et al. (2011). Regional climate and hydrologic change in the northern US Rockies and Pacific Northwest: Internally consistent projections of future climate for resource management. Seattle: Climate Impacts Group, University of Washington

College of the Environment. http://cses.washington.edu/picea/USFS/pub/Littell_etal_2010. 5 Dec 2016.

Maurer, E. P., Brekke, L., Pruitt, T., & Duffy, P. B. (2007). Fine-resolution climate projections enhance regional climate change impact studies. *Eos Transactions of the American Geophysical Union, 88*(47), 504–504.

Meehl, G. A., Covey, C., Delworth, T., et al. (2007). The WCRP CMIP3 multimodel data set. *Bulletin of the American Meteorological Society, 88*(9), 1383–1394.

Nakićenović, N., Davidson, O., Davis, G., et al. (2000). *Special report on emissions scenarios: A special report of working group III of the intergovernmental panel on climate change.* Cambridge: Cambridge University Press.

R Core Team. (2016). *R: A language and environment for statistical computing.* R Foundation for Statistical Computing. Vienna, Austria. Url: https://www.R-project.org/

U.S. Department of Agriculture, Forest Service. 2012. Future of America's forests and rangelands: Forest Service 2010 Resources Planning Act Assessment (General Technical Report WO-87). Washington, DC: U.S. Forest Service.

van Vuuren, D. P., Edmonds, J., Kainuma, M., et al. (2011). The representative concentration pathways: An overview. *Climatic Change, 109*(1-2), 5–31.

Walsh, J., Wuebbles, D., Hayhoe, K., et al. (2014). Our changing climate. In J. M. Melillo, T. C. Richmond, & G. W. Yohe (Eds.), *Climate change impacts in the United States: The third National Climate Assessment* (pp. 19–67). Washington, DC: U.S. Global Change Research Program.

Chapter 3
Effects of Climate Change on Snowpack, Glaciers, and Water Resources in the Northern Rockies

Charles H. Luce

Abstract Many of the effects of climate change on ecosystems will be mediated through changes in hydrology. Decreasing snowpack and declining summer flows with warming will alter timing and availability of water supply, affecting agricultural, municipal, and public uses in and downstream from national forests. Declining summer low flows will affect water availability during late summer, the period of peak demand for irrigation and power supply. Increased magnitude of peak streamflows will damage roads near perennial streams, ranging from minor erosion to extensive damage, thus affecting public safety, access for resource management, water quality, and aquatic habitat.

Primary adaptation strategies to address changing hydrology in the Northern Rockies include restoring the function of watersheds, connecting floodplains, reducing drainage efficiency, maximizing valley storage, and reducing hazardous fuels. Tactics include adding wood to streams, restoring American beaver populations, modifying livestock management, and reducing surface fuels and forest stand densities. Primary strategies for infrastructure include increasing the resilience of stream crossings, culverts, and bridges to higher peak flows and facilitating response to higher peak flows by reducing the road system and disconnecting roads from streams. Tactics include installing higher capacity culverts, and decommissioning roads or converting them to alternative uses. Erosion potential to protect water quality can be addressed by reducing hazardous fuels in dry forests, reducing non-fire disturbances, and using road management practices that prevent erosion.

Keywords Climate change • Streamflow • Snowpack • Glaciers • Adaptation

C.H. Luce (✉)
U.S. Forest Service, Rocky Mountain Research Station, Boise, ID, USA
e-mail: cluce@fs.fed.us

3.1 Introduction

The effects of climate change on ecosystems will be mediated through changes in hydrology. Changes in snow accumulation and melt have been documented with recent warming across the western United States (Service 2004; Barnett et al. 2005), and these changes affect when water is available for both forests and fish. In the West, changes in summer atmospheric circulation patterns may alter the ability of summer precipitation to ameliorate summer drought and dampen wildfire spread (IPCC 2013). Fish will be affected by both lower low flows with earlier snowmelt and higher midwinter floods caused by rain-on-snow events. Reduced snowpacks and declining summer water supplies will also affect municipal and agricultural uses, as approximately 70% of the water supply in the western United States is tied to mountain snowpacks (Service 2004). This chapter describes mechanisms of hydrologic change with climate warming in the Northern Rocky Mountains, with specific discussions on effects on snowpack and glaciers, streamflow, and drought, and consideration of variations and uncertainties in potential effects across the region.

3.2 Mechanisms for Climate Change Effects on Hydrology

The hydrological consequences of warmer temperatures include less snowpack and greater evaporative demand from the atmosphere. Snowpack depth, extent, and duration are expected to decrease with less precipitation falling as snow (Pierce et al. 2008), and earlier melt (Luce et al. 2014). However, the degree of change expected as a result of warming varies across the landscape as a function of temperature (Luce et al. 2014). Places that are warm (near the melting point of snow) are expected to be more sensitive than places where temperatures remain below freezing throughout much of the winter (Woods 2009). In the coldest locations, snowpack may increase with increasing winter precipitation under changing climate (Hamlet et al. 2013).

The relationship of evapotranspiration to a warming climate is complex (Roderick et al. 2014). Warmer air can hold more water, and thus, even if relative humidity stays constant, there will be an increase in vapor pressure deficit (the difference between the actual water content of the air and water content at saturation). Increased vapor pressure deficit creates a water vapor gradient between leaves and the atmosphere that can draw moisture out of leaves. Evaporation, however, is an energy intensive process, and there is only so much additional energy that will be available for evaporation. Leaves also hold moisture more tightly when conditions become dry. Thus, most of the energy from increased longwave radiation with climate change will likely result in warming rather than increased evaporation (Roderick et al. 2015).

Precipitation has a more direct impact on hydrologic processes than temperature, but potential changes in precipitation in a warming climate are more uncertain (Blöschl and Montanari 2010; IPCC 2013). On average, across many global climate models (GCMs), precipitation is expected to increase slightly in the Northern Rockies region, but the range in projections is quite large (on the order of +30 to −20%; see Chap. 2). Because of this uncertainty in precipitation, the general approach in this and other analyses is to use an ensemble average (i.e., average across many GCMs) for precipitation. The chapter also includes discussion on which processes or hydrologic outcomes are most uncertain and where.

Although there is substantial quantitative disagreement among GCMs in projected precipitation behavior, there is some agreement on the general physical mechanisms behind precipitation change. Dynamic drivers of precipitation change include changes in global circulation patterns (e.g., the Hadley cell extent) and changes in mid-latitude eddies. Changes in teleconnection patterns, for example the North American Monsoon System, would also fall into this category. Thermodynamic changes occur with warming because a warmer atmosphere can hold more water (Held and Soden 2006). According to a non-linear Clausius-Clapeyron relationship (saturation vapor pressure versus temperature), rough expectations for precipitation change are on the order of a 7% increase in precipitation per 1 °C of temperature change. There are, however, other physical limits to energy driving the cycling of water in the atmosphere, leading to lesser estimates on the order of 1.6% per 1 °C of temperature change at the global scale, with individual grid cells being less or potentially negatively affected, particularly over land (Roderick et al. 2014). Different approaches to scaling the thermodynamic contribution to precipitation is one of the reasons for differences among GCM projections, although differences in the dynamic process modeling can be significant as well.

One key outcome of thermodynamically-driven changes is that when precipitation happens, it is expected to fall with greater intensity. In turn, there will likely be longer-duration dry spells between precipitation events. These dry spells may determine drought duration in locations where summer precipitation is an important component of the summer water budget (Luce et al. 2016), such as much of eastern Montana, and low elevations in western Montana. Much of the Northern Rockies region has a substantially wetter May and June than July and August, and in some cases, May through June precipitation is on a par with or exceeds the winter snowpack contribution to the annual water budget. In these locations, snowpack changes may have fewer consequences than changes in summer precipitation, which can be an important determinant of the severity of summer drought and the fire season (Abatzoglou and Kolden 2013).

Climate change may also lead to changes in orographic enhancement of precipitation over mountain areas in the Pacific Northwest. Historical changes in westerly windflows have led to a decrease in the enhancement of winter precipitation by orographic lifting over mountain ranges (Luce et al. 2013), raising the important question of whether such a pattern may continue into the future. Westerly winds across the Pacific Northwest are strongly correlated with precipitation in mountainous areas, but valley precipitation is not, nor is precipitation in much of eastern Montana. However, because precipitation falls mostly in mountain areas where streamflows originate, the potential for future changes in orographic enhancement of precipitation are important to consider. The variable infiltration capacity (VIC) model simulations detailed later in this chapter do not include this effect, so for purposes of general discussion, it can be considered an additional source of uncertainty for precipitation.

3.3 Effects of Climate Change on Snowpack and Glaciers

Snowpack has declined across the western United States over the last few decades (Mote et al. 2005; Regonda et al. 2005; Pierce et al. 2008). Although earlier work attributed declines primarily to warming temperatures, the interior parts of the Northern Rockies are cold enough to be relatively insensitive to warming and strongly sensitive to precipitation variation (Mote 2006; Luce et al. 2014). Consequently, interior snowpacks have likely responded primarily to reduced precipitation (Luce et al. 2013). In contrast, the low-elevation mountains of northern Idaho, the westernmost mountains in the region, are heavily influenced by a maritime snow climate (Armstrong and Armstrong 1987; Mock and Birkeland 2000), and are still sensitive to temperature variability, particularly with respect to snow durability (Luce et al. 2014) (Fig. 3.1).

Glaciers are well-known features in the Northern Rockies, with a large number located in and near Glacier National Park and in the Wind River, Absaroka, and Beartooth ranges in and near Yellowstone National Park. Significant changes have been noted in the glaciers of Glacier National Park over the course of the twentieth century (Fagre 2007), with the Grinnell Glacier having approximately 10% of the ice that it had in 1850 (Fig. 3.2). Declines have also been seen in glaciers of the Wind River Range over the twentieth century (Marston et al. 1991).

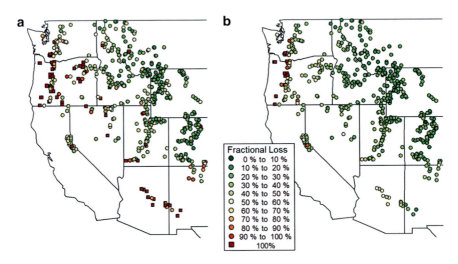

Fig. 3.1 Estimated loss of (**a**) April 1 snow water equivalent and (**b**) mean snow residence time as related to warming of 3 °C (From Luce et al. 2014)

Fig. 3.2 Oblique view of Grinnell Glacier taken from the summit of Mount Gould, Glacier National Park (After Fagre 2005)

Estimating future changes in glaciers is complex (Hall and Fagre 2003), but empirical analyses suggest a brief future for them, with many glaciers becoming fragmented or disappearing by the 2030s. Increasing temperatures yield a rising equilibrium line altitude (ELA), decreasing the effective contributing area for each glacier as warming progresses. A 3 °C warming can translate into 300–500 m of elevation rise in snow-rain partitioning. Unfortunately, for simplicity's sake, those changes do not directly equate to shift in ELA, which depends on the geometry and topography of the contributing cirque.

Temperate alpine glaciers are also sensitive to precipitation variations (McCabe and Fountain 1995). Westerlies and their contribution to winter precipitation have changed over the Glacier National Park region since the 1940s, and April 1 snow water equivalent at these elevations and latitudes is relatively insensitive to temperature. However, it is important to note that this area receives significant spring and summer precipitation, and changing summer temperatures affect both the melt rate and additional summertime mass contributions (new snow) in these glaciers. Thus, summer temperature is a strong predictor of their behavior, and regardless of changes in precipitation, significant reduction in area of glaciers is expected by the end of the twenty-first century (Hall and Fagre 2003).

3.4 Effects of Climate Change on Streamflow

Several commonly-used metrics are available to determine potential effects of climate change on streamflow. Annual yield, summer low flows, and center of runoff timing are all important metrics with respect to water supply. Irrigation water for crops and urban landscapes is typically needed in summer months, and these metrics are most relevant to surface water supplies (as opposed to groundwater supplies). For summer low flows, metrics include the mean summer yield (June through September), and the minimum weekly flow with a 10-year recurrence probability (7Q10). Center of runoff timing refers to the timing of water supply, and shifts in runoff earlier in the winter or spring disconnect streamflow timing from water supply needs.

Peakflows are important to fishes and infrastructure. Scouring flows can damage eggs in fish redds if they occur while the eggs are in the gravel or during alevin emergence (Montgomery et al. 1996; DeVries 1997; Goode et al. 2013). Winter peakflows can affect fall spawning fish (salmon Chinook [*Oncorhynchus tshawytscha*], bull trout [*Salvelinus confluentus*], brook trout [*S. fontinalis*]), whereas spring peakflows affect spring spawning cutthroat trout (*O. clarkii*), steelhead (*O. mykiss irideus*), and resident rainbow trout (*O. mykiss*) (Wenger et al. 2011a, b). Spring peakflows associated with the annual snowmelt pulse are typically muted in magnitude in comparison to winter rain-on-snow events, because the rain-on-snow events tend to affect much larger fractions of a basin at a time. Thus, a shift to more mid-winter flood events can yield greater peakflow magnitudes, which can threaten infrastructure such as roads, recreation sites, or water management facilities (e.g., diversions and dams).

Changes in streamflow metrics in recent decades have been documented in some of the western and southern basins of the Northern Rockies region. Earlier runoff timing was noted by Cayan et al. (2001) and Stewart et al. (2005), and declining annual streamflows were noted by Luce and Holden (2009) and Clark (2010). Declining low flows (7Q10), associated primarily with declining precipitation, have also been seen in the western half of the Northern Rockies (Kormos et al. 2016).

3.4.1 Future Streamflow Projections

Future streamflow projections for the Northern Rockies region were produced using the VIC model (Liang et al. 1994) (database for the western U.S. available online: https://cig.uw.edu/datasets/wus/). Climate projections were based on Coupled Model Intercomparison Project Phase 3 (CMIP3) GCM runs, the full details of which are discussed in Littell et al. (2011). The gridded data were used to estimate streamflow by using area-weighted averages of runoff from each VIC grid cell within a given basin, following the methods of Wenger et al. (2010), to accumulate flow and validate. Streamflow metrics were calculated for stream segments in the NHD + V2 stream segments (http://www.fs.fed.us/rm/boise/AWAE/projects/modeled_stream_flow_metrics.shtml). Although calculations were made for all 6th-digit Hydrologic Units in the Northern Rockies, only the western half of the region is shown in Figs. 3.3, 3.4, and 3.5, because trimming the domain allowed for better display of information. Fortunately, patterns in the easternmost portions of the figures are similar to those in the eastern portion of the Northern Rockies.

Projections indicate that mean annual streamflow may increase in the western and southern portions of the Northern Rockies, with smaller changes in eastern Montana (Fig. 3.3). There was substantial uncertainty (i.e., substantial differences among GCM projections) for annual water yield projections in higher mountains of northern Idaho and northwestern Montana. Overall, changes in the ensemble mean for annual water yield are comparable to ensemble mean changes in precipitation.

3 Effects of Climate Change on Snowpack, Glaciers, and Water Resources...

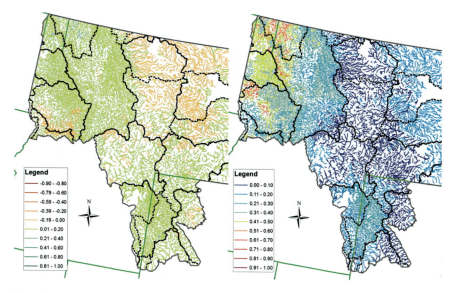

Fig. 3.3 Projections for fractional change in mean annual flow (ensemble mean) for the 2080s compared to 1977–2006

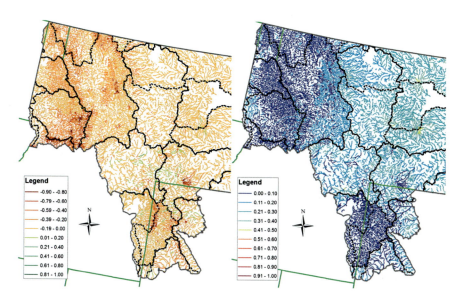

Fig. 3.4 Projections for fractional change in mean summer flow (June–September) for the 2080s (ensemble mean) compared to 1977–2006

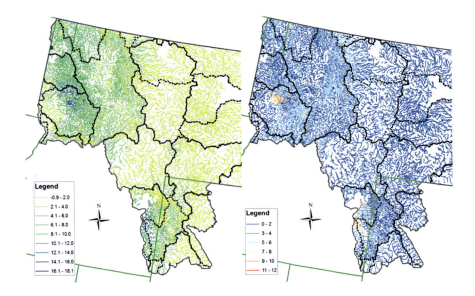

Fig. 3.5 Projections for number of days in winter that exceed the 95th percentile flow in each year (ensemble mean), an indicator of when floods are likely to happen, for the 2080s compared to 1977–2006. The value of this metric can take on values between 0 and 18.25

Although projections indicate that annual flow may increase, summer low flows are expected to decrease (Fig. 3.4). Uncertainty (i.e., differences among GCM projections) is low compared to the magnitude of projected changes, particularly in mountain areas. In general, areas showing more pronounced change in low flows show a larger shift in timing of flows (on the order of 2 months; data not shown), with greater changes in mountains with higher precipitation. The primary mechanism expected to drive lower low flows is reduced snowpack in winter, leading to less stored water.

Summer wet portions of the region are more likely to see low flows affected by summer precipitation patterns. Shifts in circulation that affect how moisture flows from the Gulf of Mexico during summer months are expected to negatively affect precipitation amounts and increase the time between precipitation events (IPCC 2013; Luce et al. 2016). Summer wet areas are also more likely to see greater losses in streamflow with increased evaporation, but it is important to recognize energy balance constraints when estimating the degree of loss (Roderick et al. 2014). This is not done in the VIC modeling, which uses only temperature outputs from GCMs without reevaluating the change in energy balance from a different hydrological formulation, leading to overestimation of loss (Milly and Dunne 2011).

Projections for changes in flood magnitude across the Northern Rockies region are substantially more uncertain and spatially heterogeneous at fine scales than those for low flows (Fig. 3.5). The shift to more midwinter rain and more rain-on-snow flooding depends strongly on the elevation range of each basin. Generally, projections indicate strong declines in flood magnitude in higher elevation basins near the crest of major mountain ranges, strong increases at middle elevations, and little change below that. However, differences among GCM projections for peak-flow magnitudes are generally as large as or larger than the expected magnitude of the change.

3.5 Adapting Water Resources and Management to Climate Change

Exploring the potential hydrological shifts in the Northern Rockies region under changing climate leads to questions about what might be done to reduce impacts on water resources. Despite the diversity in topography, geology, watershed configurations, and ecosystems across the region, the dominant climate change sensitivities and adaptation responses identified by land managers were generally consistent across the region. Most adaptation strategies and tactics were directed toward affecting downstream water availability and consequences of hydrologic drought, as little can be done to alleviate some of the more direct consequences of shifting precipitation, snowpack timing, and temperature changes to forests during drought conditions (e.g., Vose et al. 2016).

Managers were concerned about the vulnerability of roads and infrastructure to flooding, which is expected to increase at mid elevations in the Northern Rockies region. National forests have thousands of kilometers of roads, mostly unpaved. Many roads were built decades ago, and were not built to today's standards. Damage to national forest roads and associated drainage systems reduces access by users and is extremely expensive to repair. Road damage often has direct and deleterious effects on aquatic habitats as well, particularly when roads are adjacent to streams (Trombulak and Frissell 2000). Resilience to higher peakflows and frequency of flooding can be increased by (1) maintaining the capacity of floodplains and riparian areas to retain water, (2) conducting a risk assessment of vulnerable roads and infrastructure, and (3) modifying infrastructure where possible (e.g., increasing culvert size, improving road drainage, and relocating vulnerable campgrounds and road segments) (Strauch et al. 2015).

Climate-induced occurrence of disturbances such as drought, wildfire, and flooding are expected to increase, thus increasing sediment yield and affecting aquatic habitat and water resource infrastructure (Goode et al. 2012). Building an information base on potential locations of and responses to disturbances will help ensure informed and timely post-disturbance decision making (Luce et al. 2012). Specifically, managers can (1) prioritize data collection based on projections of future drought, (2) collect pre-disturbance data on water resources, and (3) develop a clearinghouse for programs related to fire and other disturbances.

Reduced overall base flows (especially in summer) are expected to alter riparian habitat, and reduce water storage in shallow aquifers in dry regions of the western United States (Perry et al. 2012). The primary adaptation strategies developed by managers in response to these expected effects were to increase natural water storage and build storage where appropriate. Specific tactics focus on (1) increasing storage with constructed wetlands, American beavers (*Castor canadensis*), and obliterated roads, and (2) considering small-scale storage in dams, retention ponds, and swales, where appropriate. In addition, it will be important to map aquifers and alluvial deposits, improve monitoring to provide feedback on water dynamics, and understand the physical and legal availability of water for aquifer recharge.

Public lands are a critical source of municipal water supplies, for which both quantity and quality are expected to decrease as snowpack declines. A key adaptation strategy is to reduce erosion potential to protect water quality, as well as prioritize municipal water supplies. Water quality can be addressed by (1) reducing hazardous fuels in dry forests to reduce the risk of crown fires, (2) reducing other types of disturbances (e.g., off-road vehicles, unregulated livestock grazing), and (3) using road management practices that reduce erosion. These tactics should be implemented primarily in high-value locations (near communities and reservoirs) on public and private lands.

Another strategy for addressing municipal water availability is to reduce water use by increasing water efficiency in federal facilities, thus strengthening the connection between the source of water on public lands and use of water downstream on public and private lands. First, it will be helpful to identify effective water-saving tactics and where they can be successfully implemented. Second, low water-use appliances can be installed at administrative sites (e.g., restrooms), and drought tolerant plants can be used for landscaping (e.g., adjacent to management unit buildings). Third, the benefit of water conservation can be communicated to users of public lands (e.g., in campgrounds). These tactics would demonstrate leadership in water conservation by the U.S. Forest Service and other agencies, and provide outreach and public relations that extend to local communities.

Acknowledgments Many thanks to Patrick Kormos and Abigail Lute for generating figures in this chapter.

References

Abatzoglou, J. T., & Kolden, C. A. (2013). Relationships between climate and macroscale area burned in the western United States. *International Journal of Wildland Fire, 22*, 1003–1020.

Armstrong, R. L., & Armstrong, B. R. (1987). Snow and avalanche climates of the western United States: A comparison of maritime, intermountain and continental conditions. Proceedings of the Davos Symposium: Avalanche formation, movement and effects. *IAHS Publication, 162*, 281–294.

Barnett, T. P., Adam, J. C., & Lettenmaier, D. P. (2005). Potential impacts of a warming climate on water availability in snow-dominated regions. *Nature, 438*, 303–309.

Blöschl, G., & Montanari, A. (2010). Climate change impacts—throwing the dice? *Hydrological Processes, 24*, 374–381.

Cayan, D. R., Dettinger, M. D., Kammerdiener, S. A., et al. (2001). Changes in the onset of spring in the Western United States. *American Meteorological Society, 82*, 399–415.

Clark, G. M. (2010). Changes in patterns of streamflow from unregulated watersheds in Idaho, Western Wyoming, and Northern Nevada. *Journal of the American Water Resources Association, 46*, 486–497. doi:10.1111/j.1752-1688.2009.00416.x.

DeVries, P. (1997). Riverine salmonid egg burial depths: Review of published data and implications for scour studies. *Canadian Journal of Fisheries and Aquatic Sciences, 54*, 1685–1698.

Fagre, D. B. (2005). Adapting to the reality of climate change at Glacier National Park, Montana, USA. Paper presented at Proceedings of the Conferencia Cambio Climatico, Bogota, Colombia, pp 14.

Fagre, D. B. (2007). Adapting to the reality of climate change at Glacier National Park, Montana, USA. In *Proceedings of The First International Conference on the Impact of Climate Change on High-Mountain Systems* (pp. 221–235). Bogota, Colombia.

Goode, J. R., Luce, C. H., & Buffington, J. M. (2012). Enhanced sediment delivery in a changing climate in semi-arid mountain basins: Implications for water resource management and aquatic habitat in the northern Rocky Mountains. *Geomorphology, 139–140*, 1–15.

Goode, J. R., Buffington, J. M., Tonina, D., et al. (2013). Potential effects of climate change on streambed scour and risks to salmonid survival in snow-dominated mountain basins. *Hydrological Processes, 27*, 750–765.

Hall, M. H., & Fagre, D. B. (2003). Modeled climate-induced glacier change in Glacier National Park, 1850–2100. *Bioscience, 53*, 131–140.

Hamlet, A. F., Elsner, M. M., Mauger, G. S., et al. (2013). An overview of the Columbia Basin climate change scenarios project: Approach, methods, and summary of key results. *Atmosphere-Ocean, 51*, 392–415.

Held, I. M., & Soden, B. J. (2006). Robust responses of the hydrological cycle to global warming. *Journal of Climate, 19*, 5686–5699.

IPCC. (2013). *Climate change 2013: The physical science basis. Contribution of working group I to the fifth assessment report of the intergovernmental panel on climate change.* Cambridge/New York: Cambridge University Press.

Kormos, P., Luce, C., Wenger, S. J., & Berghuijs, W. R. (2016). Trends and sensitivities of low streamflow extremes to discharge timing and magnitude in Pacific Northwest mountain streams. *Water Resources Research, 52*, 4990–5007.

Liang, X., Lettenmaier, D. P., Wood, E. F., & Burges, S. J. (1994). A simple hydrologically based model of land surface water and energy fluxes for general circulation models. *Journal of Geophysical Research, 99*, 14415–14428.

Littell, J. S., Elsner, M. M., Mauger, G., et al. (2011). *Regional climate and hydrologic change in the northern US Rockies and Pacific Northwest: Internally consistent projections of future climate for resource management*. Seattle: Climate Impacts Group, University of Washington College of the Environment. http://cses.washington.edu/picea/USFS/pub/Littell_etal_2010. 5 Dec 2016.

Luce, C. H., & Holden, Z. A. (2009). Declining annual streamflow distributions in the Pacific Northwest United States, 1948–2006. *Geophysical Research Letters, 36*, L16401.

Luce, C., Morgan, P., Dwire, K., Isaak, D., Holden, Z., & Rieman, B. (2012). *Climate change, forests, fire, water, and fish: Building resilient landscapes, streams, and managers*. General Technical Report RMRS-GTR-290, 207 pp., USDA Forest Service, Fort Collins, CO.

Luce, C. H., Abatzoglou, J. T., & Holden, Z. A. (2013). The missing mountain water: Slower westerlies decrease orographic enhancement in the Pacific Northwest USA. *Science, 342*, 1360–1364.

Luce, C. H., Lopez-Burgos, V., & Holden, Z. (2014). Sensitivity of snowpack storage to precipitation and temperature using spatial and temporal analog models. *Water Resources Research, 50*, 9447–9462.

Luce, C. H., Vose, J. M., Pederson, N., et al. (2016). Contributing factors for drought in United States forest ecosystems under projected future climates and their uncertainty. *Forest Ecology and Management, 380*, 299–308.

Marston, R. A., Pochop, L. O., Kerr, G. L., et al. (1991). Recent glacier changes in the Wind River Range, Wyoming. *Physical Geography, 12*, 115–123.

McCabe, G. J., & Fountain, A. G. (1995). Relations between atmospheric circulation and mass balance of South Cascade Glacier, Washington, USA. *Arctic and Alpine Research, 27*, 226–233.

Milly, P., & Dunne, K. A. (2011). On the hydrologic adjustment of climate-model projections: The potential pitfall of potential evapotranspiration. *Earth Interactions, 15*, 1–14.

Mock, C. J., & Birkeland, K. W. (2000). Snow avalanche climatology of the western United States mountain ranges. *Bulletin of the American Meteorological Society, 81*, 2367.

Montgomery, D. R., Buffington, J. M., & Peterson, N. P. (1996). Stream-bed scour, egg burial depths, and the influence of salmonid spawning on bed surface mobility and embryo survival. *Canadian Journal of Fisheries and Aquatic Sciences, 53*, 1061–1070.

Mote, P. W. (2006). Climate-driven variability and trends in mountain snowpack in western North America. *Journal of Climate, 19*, 6209–6220.

Mote, P. W., Hamlet, A. F., Clark, M. P., & Lettenmaier, D. P. (2005). Declining mountain snowpack in western North America. *Bulletin of the American Meteorological Society, 86*, 39–49.

Perry, L. G., Andersen, D. C., Reynolds, L. V., et al. (2012). Vulnerability of riparian ecosystems to elevated CO_2 and climate change in arid and semiarid western North America. *Global Change Biology, 18*, 821–842.

Pierce, D. W., Barnett, T. P., Hidalgo, H. G., et al. (2008). Attribution of declining western U.S. snowpack to human effects. *Journal of Climate, 21*, 6425–6444.

Regonda, S., Rajagopalan, B., Clark, M., & Pitlick, J. (2005). Seasonal cycle shifts in hydroclimatology over the western United States. *Journal of Climate, 18*, 372–384.

Roderick, M. L., Sun, F., Lim, W. H., & Farquhar, G. D. (2014). A general framework for understanding the response of the water cycle to global warming over land and ocean. *Hydrology and Earth System Sciences, 18*, 1575–1589.

Roderick, M. L., Greve, P., & Farquhar, G. D. (2015). On the assessment of aridity with changes in atmospheric CO_2. *Water Resources Research, 51*, 5450–5463.

Service R. F. (2004). As the West goes dry. *Science, 303*, 1124–1127.

Stewart, I. T., Cayan, D. R., & Dettinger, M. D. (2005). Changes toward earlier streamflow timing across western North America. *Journal of Climate, 18*, 1136–1155.

Strauch, R. L., Raymond, C. L., Rochefort, R. M., et al. (2015). Adapting transportation to climate change on federal lands in Washington State, USA. *Climatic Change, 130*, 185–199.

Trombulak, S. C., & Frissell, C. A. (2000). Review of ecological effects of roads on terrestrial and aquatic communities. *Conservation Biology, 14*, 18–30.

Vose, J. M., Clark, J. S., Luce, C. H., & Patel-Weynand, T. (Eds.). (2016). *Effects of drought on forests and rangelands in the United States: A comprehensive science synthesis* (General Technical Report WO-93b). Washington, DC: U.S. Forest Service, Washington Office.

Wenger, S. J., Luce, C. H., Hamlet, A. F., et al. (2010). Macroscale hydrologic modeling of ecologically relevant flow metrics. *Water Resources Research, 46*, W09513.

Wenger, S. J., Isaak, D. J., Dunham, J. B., et al. (2011a). Role of climate and invasive species in structuring trout distributions in the interior Columbia River Basin, USA. *Canadian Journal of Fisheries and Aquatic Sciences, 68*, 988–1008.

Wenger, S. J., Isaak, D. J., Luce, C. H., et al. (2011b). Flow regime, temperature, and biotic interactions drive differential declines of trout species under climate change. *Proceedings of the National Academy of Sciences, 108*, 14175–14180.

Woods, R. A. (2009). Analytical model of seasonal climate impacts on snow hydrology: Continuous snowpacks. *Advances in Water Resources, 32*, 1465–1481.

Chapter 4
Effects of Climate Change on Cold-Water Fish in the Northern Rockies

Michael K. Young, Daniel J. Isaak, Scott Spaulding, Cameron A. Thomas, Scott A. Barndt, Matthew C. Groce, Dona Horan, and David E. Nagel

Abstract Decreased snowpack with climate warming will shift the timing of peak streamflows, decrease summer low flows, and in combination with higher air temperature, increase stream temperatures, all of which will reduce the vigor of cold-water fish species. Abundance and distribution of cutthroat trout and especially bull trout will be greatly reduced, although effects will differ by location as a function of both stream temperature and competition from non-native fish species. Increased wildfire will add sediment to streams, increase peak flows and channel scouring, and raise stream temperature by removing vegetation.

Primary strategies to address climate change threats to cold-water fish species include maintaining or restoring functionality of channels and floodplains to retain (cool) water and buffer against future changes, decreasing fragmentation of stream networks so aquatic organisms can access similar habitats, and developing wildfire use plans that address sediment inputs and road failures. Adaptation tactics include using watershed analysis to develop integrated actions for vegetation and hydrology, protecting groundwater and springs, restoring riparian areas and American beaver populations to maintain summer baseflows, reconnecting and increasing off-channel habitat and refugia, identifying and improving stream crossings that impede fish movement, decreasing road connectivity, and revegetating burned areas

---

M.K. Young (✉)
U.S. Forest Service, Rocky Mountain Research Station, Missoula, MT, USA
e-mail: mkyoung@fs.fed.us

D.J. Isaak • M.C. Groce • D. Horan • D.E. Nagel
U.S. Forest Service, Rocky Mountain Research Station, Boise, ID, USA
e-mail: disaak@fs.fed.us; matthewcgroce@fs.fed.us; dhoran@fs.fed.us; dnagel@fs.fed.us

S. Spaulding • C.A. Thomas
U.S. Forest Service, Northern Region, Missoula, MT, USA
e-mail: scottspaulding@fs.fed.us; cathomas@fs.fed.us

S.A. Barndt
U.S. Forest Service, Gallatin National Forest, Bozeman, MT, USA
e-mail: sbarndt@fs.fed.us

to store sediment and maintain channel geomorphology. Removing non-native fish species and reducing their access to cold-water habitat reduces competition with native fish species.

Keywords Bull trout • Cutthroat trout • Occupancy modeling • Refugia • Water temperature

4.1 Introduction

Climate change is expected to alter aquatic ecosystems throughout the Northern Rocky Mountains. Prominent *direct changes* will include warmer stream temperatures, lower snowpack, earlier peak flows, reduced and more protracted summer baseflows, greater flow intermittence (Chap. 3), and downhill shifts in perennial channel initiation. In addition, *indirect changes* may be caused by the altered frequency and magnitude of natural disturbances. Because the fish, amphibians, crayfish, mussels, and aquatic macroinvertebrates inhabiting freshwater environments are ectotherms, water temperature dictates their metabolic rates and most aspects of their life history, including growth, migration, reproduction, and mortality. The changes in water temperature and other hydrologic characteristics associated with climate change are expected to have profound effects on their abundance and distribution.

The effects of climate change on aquatic species have been reviewed for the Pacific Northwest (Mote et al. 2003; ISAB 2007; Mantua et al. 2010; Rieman and Isaak 2010; Isaak et al. 2012a, b; Mantua and Raymond 2014) and elsewhere in the western United States (Poff et al. 2002; Ficke et al. 2007; Schindler et al. 2008 Furniss et al. 2010, 2013; Luce et al. 2012). However, empirically based, spatially explicit, and accurate projections of climate change effects on aquatic organisms are needed for scientific assessments and applications across broad geographic regions.

We developed high-resolution scenarios for stream temperature and streamflow, translating outputs from global climate models (GCMs) to habitat factors for stream reaches (Isaak et al. 2015). Scenarios were coupled with species distribution data crowd-sourced from the peer-reviewed literature and agency reports to develop species distribution models for current relationships between climate and fish species. The models were used to project probability of species habitat occupancy in streams throughout the Northern Rockies region.

We focused on climate vulnerabilities, current distribution, and projected distribution of two native salmonid species, bull trout (*Salvelinus confluentus*) and cutthroat trout (*Oncorhynchus clarkii*), which have ecological and cultural value to society and are sensitive to warm stream temperature (Eby et al. 2014). Inferences emphasized suitable habitat for juveniles of each species, because they are more thermally constrained than adults. We also evaluated the effects of nonnative species—brook trout (*S. fontinalis*), brown trout (*Salmo trutta*), and rainbow trout (*Oncorhynchus mykiss*) (the latter native to a portion of the analysis area)—on current and future habitats for native species. Isaak et al. (2015) and the Climate Shield website (http://www.fs.fed.us/rm/boise/AWAE/projects/ClimateShield.html) contain additional details on the context, framework, and databases used in this assessment.

4.2 Analytical Approach

4.2.1 Assessment Area

The assessment includes streams in national forests and national parks encompassed by the U.S. Forest Service (USFS) Northern Region (see Chap. 1). Geospatial data for the 1:100,000-scale National Hydrography Dataset (NHD)-Plus were downloaded from the Horizons Systems website (http://www.horizon-systems.com/NHDPlus/index.php, Cooter et al. 2010) to delineate a stream network, then filtered by minimum flow and maximum stream slope criteria. Summer flow values projected by the Variable Infiltration Capacity hydrologic model (VIC; Wenger et al. 2010) were obtained from the Western United States Flow Metrics website (http://www.fs.fed.us/rm/boise/AWAE/projects/modeled_stream_flow_metrics.shtml) and linked to individual stream reaches.

Stream reaches with summer flows <0.0057 $m^3 s^{-1}$, approximating a wetted width of 1 m (Peterson et al. 2013b), or with slopes >15%, were removed because they are unoccupied or support low numbers of fish (Isaak et al. 2015). Steep slopes occur at the top of drainage networks where fish populations are more vulnerable to disturbances (e.g., debris flows after wildfire) that can cause extirpations (Bozek and Young 1994; Miller et al. 2003). Thus, the 183,036-km stream network used as baseline habitat probably overestimates potential habitat, but the current resolution of analytical tools and data prevent further refinement.

4.2.2 Climate Change Scenarios

The following average summer streamflows were available from the flow metrics website: baseline period (1970–1999, hereafter 1980s) and two future periods (2030–2059, hereafter 2040s; 2070–2099, hereafter 2080s) associated with the A1B (moderate) emission scenario. An ensemble of 10 GCMs that best represented historical trends in air temperatures and precipitation for the northwestern United States during the twentieth century was used for future projections (Table 4.1). The A1B scenario used here is similar to the RCP 6.0 scenario from Coupled Model Intercomparison Project 5 simulations (see Chap. 2).

Geospatial data for August mean stream temperature were downloaded for the same A1B trajectory and climate periods from the NorWeST website and linked to the stream hydrology layer (www.fs.fed.us/rm/boise/AWAE/projects/NorWeST.html).

Table 4.1 Projected changes in mean August air temperature, stream temperature, and streamflow for major river basins in the Northern Rockies

NorWeST unit [a]	2040s (2030–2059)			2080s (2070–2099)		
	Air temperature change (°C)[b]	Stream flow change (%)[b]	Stream temperature change (°C)[c]	Air temperature change (°C)	Stream flow change (%)	Stream temperature change (°C)
Yellowstone	2.81	−4.1	1.01	5.08	−5.4	1.81
Clearwater	3.17	−23.9	1.62	5.45	−34.2	2.78
Spokoot	3.05	−20.1	1.27	5.33	−31.5	2.19
Upper Missouri	3.25	−14.9	1.17	5.47	−21.3	1.94
Marias-Missouri	2.91	−10.0	0.75	5.30	−18.7	1.37

Projections are based on output from an ensemble of 10 global climate models for the A1B emission scenario. See text for more details on modeling and data sources (Isaak et al. 2010; Mote and Salathé 2010; Wenger et al. 2010; Hamlet et al. 2013; Luce et al. 2014)
[a]Unit boundaries described at the NorWeST website
[b]Changes in air temperature and streamflow expressed relative to 1980s (1970–1999) baseline climate period
[c]Stream temperature change accounts for differential sensitivity to climate forcing within and among river basins (Luce et al. 2014; NorWeST website)

Then, the NorWeST scenarios were developed using spatial statistical network models (Ver Hoef et al. 2006; Isaak et al. 2010) applied to 11,703 summers of monitoring data at 5461 stream sites monitored with digital sensors from 1993 to 2011. Additional rationale and criteria associated with climate scenarios and stream temperature modeling are discussed in Isaak et al. (2015).

4.2.3 Fish Species

Bull trout in the Northern Rockies are largely from an inland lineage (Ardren et al. 2011), and may express migratory or resident life histories. Migratory fish travel long distances as subadults to more productive habitats, achieving larger sizes and greater fecundity as adults before returning to natal habitats to spawn. Resident fish remain in natal habitats and mature at smaller sizes, although often at the same age as migratory adults. Adults spawn and juveniles rear almost exclusively in streams with average summer water temperature <12 °C and flow >0.034 m^{3}s^{-1} (Rieman et al. 2007; Isaak et al. 2010). This species has declined substantially compared to its historical distribution because of water development and habitat degradation (leading to higher water temperatures and lower habitat complexity), human-created migration barriers, harvest by anglers, and interactions with nonnative fishes (Rieman et al. 1997). Nonnative brook trout, brown trout, and lake trout (*Salvelinus namaycush*) compete with or prey on bull trout (Martinez et al. 2009; Al-Chokhachy et al. 2016), or cause wasted reproductive opportunities (Kanda et al. 2002). Bull trout are listed as threatened under the U.S. Endangered Species Act (ESA) (USFWS 2015).

Cutthroat trout were represented by two subspecies. Westslope cutthroat trout (*O. c. lewisi*) have a complicated phylogenetic history, with a northern/eastern lineage that occupied and colonized river basins influenced by glaciation, and a suite of southern/western lineages in basins never influenced by glaciation (M. Young, unpublished data). These fish also exhibit resident and migratory life histories, and may spawn and rear in smaller (<70 cm wide) and warmer (up to 14 °C) streams than do bull trout (Peterson et al. 2013a, b; Isaak et al. 2015). Yellowstone cutthroat trout (*O. c. bouvieri*) has an unresolved distribution because of its complex geohydrologic history associated with the Bonneville Basin. Life histories and spawning and juvenile habitats are presumed to be the same as for westslope cutthroat trout.

Distributions of both subspecies have declined >50% in response to the same stressors affecting bull trout (Shepard et al. 2005; Gresswell 2011), although cutthroat trout appears to occupy a larger proportion of its historical habitat than bull trout. Both subspecies of cutthroat trout have been petitioned under the ESA, but found not warranted for listing. Brook trout have replaced cutthroat trout in many areas, especially in the upper Missouri River basin (Shepard et al. 1997), facilitated by the distribution of low-gradient alluvial valleys (Benjamin et al. 2007; Wenger et al. 2011a). Where rainbow trout have been introduced outside their native range, introgressive hybridization occurs with cutthroat at lower elevations and in warmer waters (Rasmussen et al. 2012; McKelvey et al. 2016b). Yellowstone cutthroat trout have been widely stocked throughout the historical range of westslope cutthroat trout (Gresswell and Varley 1988), resulting in hybridization (McKelvey et al. 2016b). Lake trout predation greatly reduced Yellowstone cutthroat populations in Yellowstone Lake in the early twenty-first century, but cutthroat trout populations have rebounded somewhat following predator control (Syslo et al. 2011).

4.2.4 Trout Distribution Models

Species distribution models were developed to predict occurrence probabilities of juvenile bull trout and cutthroat trout; juvenile presence is indicative of natal habitat and a locally reproducing population (Rieman and McIntyre 1995; Dunham et al. 2002). Juvenile distributions are more restricted than those of adults, especially with respect to temperature (Elliott 1994). Juvenile bull trout are rarely found where mean summer temperatures exceed 12 °C (Dunham et al. 2003; Isaak et al. 2010), whereas some adult bull trout occupy habitats as much as 5–10 °C warmer (Howell et al. 2010). Similar patterns exist for cutthroat trout (Schrank et al. 2003; Peterson et al. 2013a), so a thermal criterion was also used to define suitable habitat for juvenile cutthroat.

A mean August stream temperature of 11 °C was chosen as the temperature criterion after cross-referencing thousands of species occurrence observations in Montana, Idaho, and Wyoming against temperature estimates from the NorWeST baseline scenario. Fish data were contributed by state and federal agencies (Isaak et al. 2015). Most native juvenile trout (90% of bull trout observations, 75% of cutthroat trout observations) occurred at sites less than 11 °C, whereas most nonnative species were rare at those sites. The thermal niche of brook trout overlapped that of the native species, but peaked at a slightly warmer temperature. Very cold temperatures reduced rainbow trout incursions and limited their introgression with cutthroat trout, particularly below 9 °C (Rasmussen et al. 2012; Young et al. 2016).

Spatially contiguous 1-km reaches of streams with temperature <11 °C were aggregated into discrete cold-water habitats (CWHs), and the fish survey data were used to assign occupancy (present, absent) by native trout juveniles and brook trout. Logistic regressions modeled the probability of native trout occupancy as a function of CWH size, stream slope, brook trout prevalence, and stream temperature. For each CWH, habitat size was represented as channel length, stream slope as the average value across all reaches, and brook trout prevalence as percentage of sample sites where they occurred.

Classification accuracy of the models at a 50% occupancy threshold was 78.1% for bull trout and 84.6% for cutthroat trout. The final logistic regression models included the four main predictor variables and some interactions among the variables. Habitat occupancy for both native trout was positively related to CWH size, but bull trout required habitats five times larger than cutthroat trout to achieve comparable probabilities of occupancy. Bull trout occupancy declined as minimum temperature warmed, whereas cutthroat trout occupancy was positively related to mean temperature. Stream slope and co-occurrence with brook trout negatively affected both species, especially in small streams. The presence of brook trout masked the apparent preference of cutthroat trout for habitats with low slopes.

The logistic regression models were applied to the full set of CWHs within the historical range of each native species in the Northern Rockies. Occupancy probabilities were calculated for a no-brook-trout scenario and a scenario in which brook trout were present at 50% of sites within each CWH. We did not include a scenario in which brook trout were present at all sites, because their prevalence rarely exceeded 50% in the large CWHs, and because not all locations were suitable for brook trout (Wenger et al. 2011a).

Species probability maps were cross-referenced with land administrative status using geospatial data from the U.S. Geological Survey Gap Analysis Program (Gergely and McKerrow 2013). The total length and percentage of CWHs and stream temperatures were summarized by jurisdiction for different climate periods. CWHs with occupancy probabilities exceeding 90% were considered potential climate refugia for native trout.

4.3 Vulnerability of Native Trout to Climate Change

4.3.1 Stream Temperature

A high level of thermal heterogeneity exists across the complex topography and elevation range of Northern Rockies streams (Fig. 4.1). Of the 183,036 km of stream habitat within the analysis area, 38% had a mean August temperature <11 °C. Most of those CWHs (86%) were in publicly administered lands, primarily (69%) in national forests. Areas with concentrations of cold streams were generally

4 Effects of Climate Change on Cold-Water Fish in the Northern Rockies

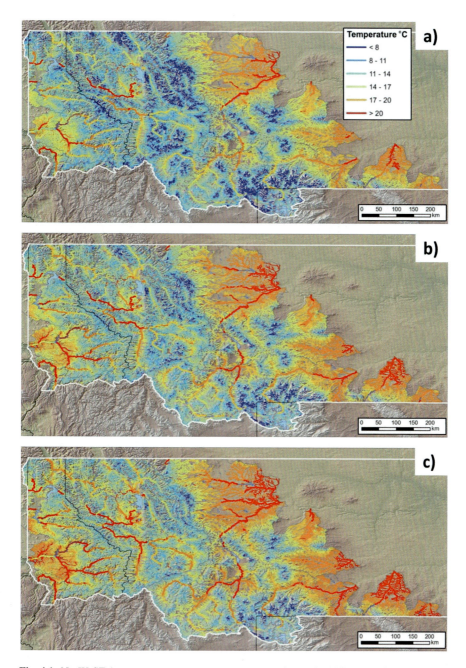

Fig. 4.1 NorWeST August mean stream temperature maps interpolated from 11,703 summers of monitoring data at 5461 unique stream sites across 183,500 km of streams in the Northern Rockies. Map panels show conditions during baseline (**a**, 1980s), moderate (**b**, 2040s), and extreme change scenarios (**c**, 2080s). See text for details on analytical methods and data sources

Table 4.2 Lengths of streams (km) in the Northern Rockies categorized by mean August stream temperature during the baseline and two future climate periods, and by land administrative status

Land status	<8 °C	8–11 °C	11–14 °C	14–17 °C	17–20 °C	>20 °C	Totals
Forest Service lands							
1980s[a]	11,416 (17.4)	36,717 (56.0)	15,034 (22.9)	1957 (3.0)	393 (0.6)	27 (0)	65,544
2040s[b]	4030 (6.3)	28,739 (44.7)	25,607 (39.8)	4976 (7.7)	716 (1.1)	194 (0.3)	64,262
2080s[b]	1547 (2.4)	20,441 (32.3)	30,660 (48.5)	8891 (14.1)	1345 (2.1)	381 (0.6)	63,265
Non-Forest Service lands							
1980s	2501 (2.1)	19,290 (16.4)	41,272 (35.1)	34,215 (29.1)	17,644 (15.0)	2571 (2.2)	117,493
2040s	915 (0.8)	9593 (8.4)	33,764 (29.5)	39,303 (34.3)	24,838 (21.7)	6148 (5.4)	114,561
2080s	407 (0.4)	5549 (4.9)	26,395 (23.3)	40,440 (35.7)	29,628 (26.1)	10,941 (9.7)	113,360

Values in parentheses are percentages of the total in the last column
[a]Stream reaches with slope <15% and VIC summer flows >0.0057 m^{3}s^{-1}
[b]Reduced network extent results from projected decreases in summer flows per Table 4.1

associated with high-elevation, steep mountain ranges in Montana, whereas such concentrations were absent from most of northern Idaho.

Mean August stream temperature was projected to increase across the Northern Rockies by an average of 1.2 °C in the 2040s and 2.0 °C in the 2080s (Table 4.1, Fig. 4.1). Increases will be disproportionately higher in the warmest streams at low elevations, and lower for the coldest streams. Differential warming occurs because cold streams are often buffered by influxes of groundwater (Luce et al. 2014). Averaged across all streams, future projections imply faster rates of warming (0.2–0.3 °C per decade) than were observed recently (0.1–0.2 °C per decade; Isaak et al. 2012a).

Based on these projections, the length of streams with temperatures <14 °C will decrease to 43,277 km in the 2040s and 27,944 km in the 2080s (Table 4.2). In both scenarios, >75% of the cold streams are in national forests. Very cold streams likely to provide habitat for bull trout and cutthroat trout originate along the Continental Divide in northern Montana, several smaller mountain ranges scattered throughout central Montana, and along the northern flank of the Beartooth Plateau (Fig. 4.1). Persistent CWHs are more isolated elsewhere.

Table 4.3 Number and length of cold-water habitats for juvenile cutthroat trout by probability of occurrence during three climate periods and two brook trout invasion scenarios across the Northern Rockies

Metric and brook trout prevalence	Period	Probability of occurrence (%)					Total
		<25	25–50	50–75	75–90	>90	
Cold-water habitat number							
0% brook trout prevalence	1980s	71	392	1314	1817	1739	5333
	2040s	41	328	1405	1505	1148	4427
	2080s	86	659	949	977	770	3441
50% brook trout prevalence	1980s	73	501	2790	1384	581	5329
	2040s	41	382	2571	1065	367	4426
	2080s	86	684	1837	673	161	3441
Cold-water habitat length (km)							
0% brook trout prevalence	1980s	432	1278	4068	7730	32,646	46,154
	2040s	126	898	3832	6034	17,964	28,856
	2080s	229	1659	2938	4151	10,459	19,436
50% brook trout prevalence	1980s	387	2344	10,320	13,201	19,348	45,602
	2040s	126	1376	8174	8772	10,306	28,756
	2080s	228	1992	6327	6289	4598	19,436

4.3.2 Cutthroat Trout

The historical range of cutthroat trout includes most of the Northern Rockies. The number of discrete CWHs for cutthroat trout in the baseline climate period was estimated to exceed 5000, encompassing over 45,000 km of streams (Table 4.3, Fig. 4.2). Over 90% of CWHs had occupancy probabilities exceeding 50%, because cutthroat require relatively small stream networks (10 km is associated with an occupancy probability of 90%; also see Peterson et al. 2013a). The largest CWHs contained a disproportionate amount of habitat most likely to be occupied; 32.6% were climate refugia, which accounted for 70.7% of the length of CWHs.

In future scenarios, the number and extent of CWHs decreased 20–60%, but even under the extreme scenario nearly 3500 potential CWHs (>19,000 km) were projected to remain. And in a few basins currently too cold for cutthroat trout (e.g., Teton River basin along the Rocky Mountain Front, streams in northern Yellowstone National Park), future warming is expected to increase their suitability. Although the presence of brook trout did not alter the number of CWHs, it did decrease the probability of cutthroat trout occupancy (Table 4.3). Sensitivity of streams to brook trout varied with local conditions, with the greatest reductions in small streams with relatively shallow slopes.

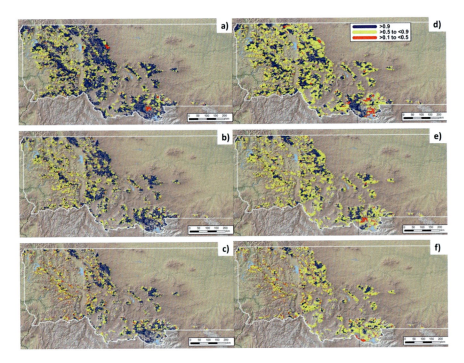

Fig. 4.2 Distribution of cold-water habitats with occupancy probabilities >0.1 for juvenile cutthroat trout during baseline (**a** & **d**, 1980s), moderate-change (**b** & **e**, 2040s), and extreme-change scenarios (**c** & **f**, 2080s). Panels **a–c** illustrate occupancy when brook trout are absent; panels **d–f** illustrate occupancy when brook trout prevalence is 50%. See text for details on analytical methods and data sources

4.3.3 Bull Trout

The number of discrete CWHs for bull trout during the baseline climate period exceeded 1800, encompassing >23,000 km (Table 4.4, Fig. 4.3). Occupancy probabilities for most bull trout CWHs were <50% because of the large stream networks required by this species (50 km is associated with occupancy probability of 90%). Only 6% of CWHs were considered climate refugia, but they were 30% of the total length of CWH. This requirement for large CWHs caused projected decreases in the number and extent of bull trout CWHs to be much higher (38–71%) than for cutthroat trout, particularly for CWHs with the highest occupancy probabilities. More than 800 CWHs representing over 7000 km were projected to remain, even in the extreme scenario.

Brook trout invasions reduced bull trout occupancy rates. These declines were more pronounced for bull trout than cutthroat trout, especially in CWHs most likely to be occupied (those with >50% occupancy probability). Fewer than 10 climate refugia for bull trout are projected to remain under any warming scenario if brook trout occupy half of each CWH. However, many large CWHs for bull trout appear

4 Effects of Climate Change on Cold-Water Fish in the Northern Rockies

Table 4.4 Number and length of cold-water habitats for juvenile bull trout by probability of occurrence during three climate periods and two brook trout invasion scenarios across the Northern Rockies

Metric and brook trout prevalence	Period	Probability of occurrence (%)					Total
		<25	25–50	50–75	75–90	>90	
Cold-water habitat number							
0% brook trout prevalence	1980s	875	534	248	92	106	1855
	2040s	664	314	98	41	32	1149
	2080s	474	274	81	24	13	866
50% brook trout prevalence	1980s	995	484	181	65	28	1753
	2040s	697	270	63	17	5	1052
	2080s	535	260	49	5	3	852
Cold-water habitat length (km)							
0% brook trout prevalence	1980s	4677	5099	4128	2601	7495	24,002
	2040s	3583	3112	1817	1237	2157	11,906
	2080s	2109	2130	1243	622	932	7035
50% brook trout prevalence	1980s	6309	6055	4365	3043	3783	23,554
	2040s	4390	3553	1916	948	656	11,462
	2080s	2525	2647	1133	246	428	6980

less susceptible to brook trout invasions (Isaak et al. 2015). CWHs with the highest bull trout occupancy probabilities during all climate periods and brook trout invasion scenarios were associated with river networks with a high number of cold streams (e.g., Whitefish River, North Fork Blackfoot River, and headwater portions of the North and Middle Forks of the Flathead River) (Figs. 4.1 and 4.3). Because of the lower elevations and warmer streams in northern Idaho, few or no climate refugia were projected to remain under either warming scenario.

4.3.4 Additional Fish Species

Native fish species other than bull trout and cutthroat trout occupy streams throughout the Northern Rockies, but were not considered priorities for this assessment because they are not expected to be as sensitive to warming temperatures as cold-water salmonids. Prairie fish in the Grassland subregion are a geographically discrete group of species that are tolerant of warm water but may be sensitive to other climate-related stressors (e.g., low water levels) (Box 4.1). Additional fish species in the Northern Rockies can be considered as candidates for the habitat occupancy-climate vulnerability approach described here.

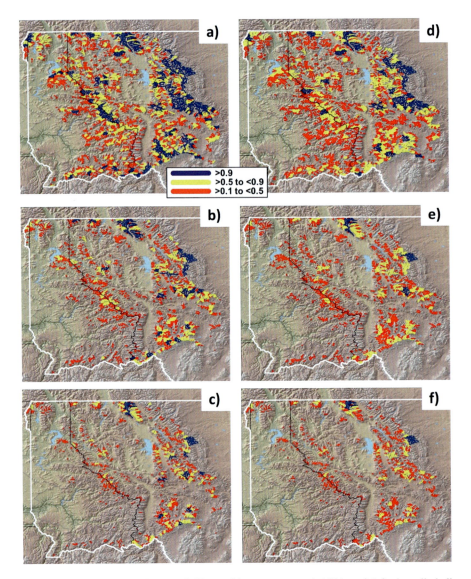

Fig. 4.3 Distribution of cold-water habitats with occupancy probabilities >0.1 for juvenile bull trout during baseline (**a** & **d**, 1980s), moderate-change (**b** & **e**, 2040s), and extreme-change scenarios (**c** & **f**, 2080s). Panels a–c illustrate occupancy when brook trout are absent; panels d–f illustrate occupancy when brook trout prevalence is 50%. See text for details on analytical methods and data sources

Box 4.1 Climate Change Effects on Fish Species in the Grassland Subregion

Several fish species are found in the Grassland subregion of the Northern Rockies. Located in the eastern portions of Custer-Gallatin National Forest and Dakota Prairie Grasslands, these species have received little scientific study and monitoring compared to cold-water salmonids and warm-water sportfish. Most prairie streams have been poorly sampled, making fish populations and aquatic habitat difficult to evaluate. Small streams constitute the majority of fish habitat, providing seasonal habitats for spawning and rearing of species favoring larger streams, rivers, and lakes.

Prairie streams are dynamic, varying between periods of floods and intermittent flows within and between years. Extirpation and recolonization of local habitats by fish species is the norm, with fish species distributed as metapopulations. Although it is typical for prairie streams to be reduced to sets of disconnected pools in some years, this pattern is more prevalent in agricultural landscapes where surface and groundwater withdrawals are common. Climate change is expected to cause greater extremes, including both severe droughts and wet intervals in dryland systems.

Projecting responses of prairie fishes to climate change is complicated by difficulty in identifying habitat preferences, because many species are habitat generalists, and interannual habitat occupancy is difficult to quantify. Prairie fish assemblages include four species guilds—northern headwaters, darter, madtom, and turbid river guilds—that are likely to differ in their vulnerability to climate change. Annual air temperature and various measures of streamflow are strong predictors of presence for the northern headwaters, darter, and madtom guilds.

The northern headwaters guild may be most vulnerable to increasing temperature, as well as to climate-related decreases in groundwater recharge. This guild includes the northern redbelly dace (*Chrosomus eos*), a sensitive species in the USFS Northern Region, which occupies small, relatively cool headwater streams. Accurate mapping of habitat types, species assemblages, and monitoring of habitat conditions will help refine potential climate change effects on habitat and species, as well as suggest appropriate management responses.

Buffering variations in flow extremes (e.g., securing instream flows or facilitating American beaver colonization where suitable habitat exists) and encouraging the presence of riparian vegetation are sound climate change adaptation options where the northern headwaters guild is present. Although

(continued)

Box 4.1 (continued)

other prairie fish guilds seem less vulnerable to changes to temperature, all are influenced by amount and timing of flow, so adaptation strategies for the northern headwater guild should also be appropriate for other guilds. All guilds are currently at risk, and may become more so if flow regimes become more variable. If migration barriers are present, it would be prudent to remove them to facilitate fish movement, while being cognizant of the potential for nonnative fish to become established. Responses of nonnative fish to climate change are uncertain, although some species (e.g., smallmouth bass [*Micropterus dolomieu*]), are expected to expand their distribution.

4.4 Applying the Assessment

The assessment described above provides spatially explicit projections of habitat occupancy in the Northern Rockies by combining ecological understanding of cutthroat trout and bull trout, species distribution data, and high-resolution projections of stream temperature and streamflow. Projections of habitat occupancy in response to anticipated climate change have several implications for future viability of native fish populations in the Northern Rockies and for conservation of these species.

Both native trout species require cold-water habitat, but their response to warmer stream temperatures will differ. Bull trout are adapted to some of the coldest freshwater environments in the Northern Hemisphere (Klemetsen et al. 2003), inhabiting variable environments with strong productivity gradients that favor migration as a life history tactic (Klemetsen 2010). Because bull trout in this region require cold water, are near the southern end of their range, and have inherently low populations in most locations (High et al. 2008), their susceptibility to range contraction in a warmer climate is unsurprising. We anticipate large reductions in their distribution in the Northern Rockies because climate refugia are relatively uncommon and dispersed, but at least some climate refugia will be retained in the future, making it more likely that bull trout will persist. The conditions favoring migratory or resident life histories may change, perhaps in uncertain ways, and it remains to be seen how to accommodate or exploit this transition in conservation practices. Research and monitoring can provide a better understanding of environmental drivers of bull trout life history.

Cutthroat trout can accommodate a broader range of thermal environments, commensurate with their evolutionary history and extensive latitudinal distribution. Their life history strategies are flexible, ranging from migratory populations that use large water bodies for growth and fecundity, to resident populations with low mobility that have been isolated for decades (Northcote 1992; Peterson et al. 2013a). The distribution of cutthroat trout is expected to decrease in the Northern Rockies, but not as much as that of bull trout. In addition, some basins currently too cold to support cutthroat trout may become suitable as the climate warms (Cooney et al. 2005; Coleman and Fausch 2007).

The degree to which nonnative salmonids displace bull trout and cutthroat trout in a warmer climate will have a major impact on long-term population viability and conservation strategies (Wenger et al. 2011a). Tolerance of cold temperature by brook trout is nearly equivalent to that of cutthroat trout, and they are especially competitive in the low-gradient environments preferred by bull trout and cutthroat trout (Wenger et al. 2011a). Large habitats (>100 km long) are less susceptible to incursions by brook trout, at least partially because they face competition from rainbow trout or brown trout (Fausch et al. 2009), species expected to shift upstream in a warmer climate (Wenger et al. 2011b; Isaak and Rieman 2013).

The USFS will play a major role in the conservation of native fish populations because most cold-water habitats in the Northern Rockies are in national forests (Table 4.2). Active management that conserves native fish is an option, because most of the cold-water habitats are outside designated wilderness areas and national parks that restrict management activities. Even under extreme warming, cold-water habitats are expected to persist in some river basins in Montana. Maintenance of these conditions is critical. In locations where climate refugia are unlikely to persist (Clearwater, Spokane, and Kootenai River basins in Idaho), active management—manipulation of habitat, fish populations, or both—may be the only way to ensure long-term persistence of native fish populations. Retaining native trout populations in some areas may require costly conservation investments, so it will be important to prioritize projects where success is likely and where benefits can be gained for other resources (e.g., riparian restoration or improved water quality).

The model projections described above are consistent with trends that have been occurring in the Northern Rockies during the last 50 years: increased air temperature, increased stream temperature, and decreased summer streamflow (Luce and Holden 2009; Isaak et al. 2010, 2012a,b; Leppi et al. 2012). This provides validation that modeled estimates of occupancy probabilities are biologically robust, facilitating a spatially explicit ranking of critical habitats. The Climate Shield fish distribution maps and databases developed in this assessment are easy to understand and access, allowing users to quantify the likely amount, distribution, and persistence of native trout habitats at multiple spatial scales (e.g., stream, river network, national forest, or region).

In general, model output suggests that environmental gradients are the primary drivers of habitat occupancy by juvenile native trout. Model projections can be improved in the future by including more local information on habitat conditions (Peterson et al. 2013a), especially the presence of barriers that influence habitat connectivity (Erős et al. 2012), and by applying spatial network models (Isaak et al. 2014). An ongoing assessment process can reduce uncertainties about distribution of aquatic species and climate change responses. Currently available data were derived from thousands of sites, but additional data would improve existing models and help develop models for additional species. Ongoing assessment and updated modeling can be combined with new surveys, such as those based on rapid and reliable environmental DNA surveys (McKelvey et al. 2016a), to provide a more accurate picture of species distribution at fine spatial scales. Such surveys will also expand our capability of assessing multiple species simultaneously.

4.5 Adapting Fish Species and Fisheries Management to Climate Change

4.5.1 Adaptation Options

Climate change adaptation for fish conservation has been reviewed extensively for western North America, including for the Northern Rockies (ISAB 2007; Rieman and Isaak 2010; Isaak et al. 2012a; Beechie et al. 2013; Luce et al. 2013; Williams et al. 2015), based on a relatively well-established set of climate sensitivities and adaptation options (Rieman et al. 2007; Mantua and Raymond 2014; Isaak et al. 2015). This provides credibility and consistency for sustainable management of fisheries in a warmer climate. This information, combined with the Northern Rockies fisheries assessment, provide the foundation for federal resource specialists to develop strategic (general, overarching) and tactical (specific, on the ground) management responses that improve the resilience of fish populations in a warmer climate.

Climate change sensitivities and adaptation options are similar among the mountainous subregions of the Northern Rockies. An exception occurs in the Eastern Rockies and Grassland subregions, where livestock grazing is a significant stressor. The Grassland subregion has no cold-water fish species and is dominated by warm-water species, many of which are nonnatives. Although some concern exists about aquatic systems in this subregion, no adaptation options were developed for fisheries in the Grassland subregion (but see Box 5.2).

Reduced snowpack is a well-documented effect of warmer temperatures in mountainous regions (Chap. 3), resulting in lower summer streamflows and warmer stream temperatures. Adaptation strategies include maintaining higher summer flows and reducing the negative effects of lower flows. On-the-ground adaptation tactics include pulsing flows from regulated streams when temperature is high, reducing water withdrawals for human uses, and securing water rights for instream flows to control overall water supply.

Increasing cold-water habitat resilience by maintaining and restoring structure and function of streams is another important adaptation strategy. Adaptation tactics include restoring channel and floodplain structure to retain cool water and riparian vegetation, and ensuring that passages for aquatic organisms are effective. These tactics can be leveraged with ongoing habitat restoration activities, especially near roads and where high streamflows are frequent. As a general principle, accelerating riparian restoration will be an effective way to improve hydrologic function and water retention. Maintaining and restoring American beaver (*Castor canadensis*) populations is also an excellent approach for water retention in mountain landscapes. Finally, road removal and relocation from locations near stream channels and floodplains can greatly improve hydrologic function.

Interactions with nonnative fish species are a significant stress for native cold-water fish in the Northern Rockies. Facilitating movement of native fish to locations with suitable stream temperatures is a primary adaptation strategy. Adaptation tactics include increasing the size of suitable habitat, modifying or removing barriers to fish passage, and documenting where groundwater inputs provide cold water. Efficacy of these tactics will be higher if native fish populations are currently healthy and nonnatives are not well established. Another important adaptation strategy is reducing nonnative fish species. Adaptation tactics include increased harvest of nonnative fish (e.g., sport fishing), manual or chemical removal of nonnatives, and excluding nonnatives with migration barriers where feasible.

Livestock grazing can damage vegetation adjacent to streams in grasslands and shrublands, predisposing aquatic systems to further degradation from warmer stream temperatures. One adaptation strategy is managing grazing to reduce damage and restore ecological and hydrologic function of riparian systems. Adaptation tactics include ensuring that standards and guidelines for water quality are adhered to and monitored, making improvements that benefit water quality and riparian shading (e.g., fencing), and reducing the presence of cattle through the retirement of vacant grazing allotments. Locations with high ecological value can be prioritized.

In a warmer climate, it is almost certain that increased wildfire occurrence will contribute to erosion and sediment delivery to streams, thus reducing water quality. Increasing resilience of vegetation to wildfire is an adaptation strategy that can help reduce the severity of fires when they occur. Hazardous fuel treatments that reduce forest stand densities and surface fuels are an adaptation tactic that is already widely used in dry forest ecosystems. Disconnecting roads from stream networks, another tactic already in practice, is especially important, because most sediment delivery following wildfire is derived from roads.

4.5.2 Principles of Climate-Smart Management

Adaptation options summarized here provide a diverse range of management responses to climate change for fisheries managers. In addition to these adaptation options, several overarching principles can help guide implementation:

- *Be strategic* — Prioritize watershed restoration to ensure that the most important work is done in the most important places. For example, climate refugia for native trout in wilderness areas may not require habitat modification to ensure persistence of those populations. Similar refugia outside wilderness can be targeted to improve habitat conditions and reduce nonnative species. Some basins are unlikely to provide suitable habitat for native trout in the future, so direct conservation investments elsewhere.

- *Implement monitoring programs* — Reduce current and future uncertainties for decision-making with strategic monitoring, then revise assessments and adaptation as needed. More data are needed for streamflow (more sites), stream temperature (annual data from sensors), and fish distributions. These data will improve knowledge of status and trends, and contribute to improved models. Monitoring efficiency is being improved with eDNA inventories of aquatic organisms (Carim et al. 2016) and inexpensive temperature and flow sensors (USEPA 2014).
- *Restore and maintain cold stream temperatures* — Many options exist: relocate roads away from streams, limit seasonal grazing, and manage riparian forest to maintain shade. In addition, take advantage of existing restoration programs to improve aquatic habitat for native fish populations.
- *Manage connectivity* — Remove obstacles to fish migration to enhance the success of migratory life history forms of native fish species, but be aware that increased connectivity can also provide access for nonnative fish species (Fausch et al. 2009). Native populations above barriers may be secure if they can adopt resident life histories, but are susceptible to extreme disturbances.
- *Remove nonnative species* — Use chemical treatments or electrofishing to remove nonnative fish species in smaller habitats, thus reducing stress on native fish populations. Control measures can be useful even if all nonnatives cannot be removed, although a migration barrier to prevent reinvasion and periodic additional controls will generally be needed to improve effectiveness.
- *Implement assisted migration* — Move native fish species from one location to another, a historically common activity in fish management, to found populations in previously fishless or formerly occupied waters. Although controversial for most taxa, assisted migration (or managed relocation) may be useful where basins are currently fishless (or contain only nonnative species in limited numbers) because of natural barriers, with the potential to be climate refugia in the future. Repeated introductions of native species may be appropriate when natural refounding is not an option, such as when populations are isolated and susceptible to periodic population crashes (Dunham et al. 2011).

Fisheries managers require a portfolio of strategic and tactical adaptation options, as described here, to address the many biogeographic circumstances they will encounter in the future. Stream habitats are already dynamic and will be even more variable in a warmer climate, undergoing both gradual and episodic changes over time. Many fish populations will adapt successfully, but others will be extirpated. Although it may not be possible to preserve all populations of all fish species in their current location, new data can inform adaptive management that targets conservation where it is most likely to succeed.

References

Al-Chokhachy, R., Schmetterling, D. A., Clancy, C., et al. (2016). Are brown trout replacing or displacing bull trout populations in a changing climate? *Canadian Journal of Fisheries and Aquatic Sciences, 73*, 1395–1404.

Ardren, W. R., DeHaan, P. W., Smith, C. T., et al. (2011). Genetic structure, evolutionary history, and conservation units of bull trout in the coterminous United States. *Transactions of the American Fisheries Society, 140*, 506–525.

Beechie, T., Imaki, H., Greene, J., et al. (2013). Restoring salmon habitat for a changing climate. *River Research and Applications, 29*, 939–960.

Benjamin, J. R., Dunham, J. B., & Dare, M. R. (2007). Invasion by nonnative brook trout in Panther Creek, Idaho: Roles of local habitat quality, biotic resistance, and connectivity to source habitats. *Transactions of the American Fisheries Society, 136*, 875–888.

Bozek, M. A., & Young, M. K. (1994). Fish mortality resulting from delayed effects of fire in the Greater Yellowstone Ecosystem. *Great Basin Naturalist, 54*, 91–95.

Carim, K. J., McKelvey, K. S., Young, M.K., et al. (2016). A protocol for collecting environmental DNA samples from streams. *U.S. Forest Service General Technical Report* (RMRS-GTR-355). Fort Collins: U.S. Forest Service, Rocky Mountain Research Station.

Coleman, M. A., & Fausch, K. D. (2007). Cold summer temperature limits recruitment of age-0 cutthroat trout in high-elevation Colorado streams. *Transactions of the American Fisheries Society., 136*, 1231–1244.

Cooney, S. J., Covich, A. P., Lukacs, P. M., et al. (2005). Modeling global warming scenarios in greenback cutthroat trout (*Oncorhynchus clarki stomias*) streams: Implications for species recovery. *Western North American Naturalist, 65*, 371–381.

Cooter, W., Rineer, J., & Bergenroth, B. (2010). A nationally consistent NHDPlus framework for identifying interstate waters: Implications for integrated assessments and interjurisdictional TMDLs. *Environmental Management, 46*, 510–524.

Dunham, J. B., Rieman, B. E., & Peterson, J. T. (2002). Patch-based models to predict species occurrence: lessons from salmonid fishes in streams. In J. M. Scott, P. J. Heglund, M. Morrison, et al. (Eds.), *Predicting species occurrences: Issues of scale and accuracy* (pp. 327–334). Covelo, CA: Island Press.

Dunham, J. B., Rieman, B. E., & Chandler, G. L. (2003). Influences of temperature and environmental variables on the distribution of bull trout within streams at the southern margin of its range. *North American Journal of Fisheries Management, 23*, 894–904.

Dunham, J. B., Gallo, K., Shively, D., et al. (2011). Assessing the feasibility of native fish reintroductions: A framework applied to threatened bull trout. *North American Journal of Fisheries Management, 31*, 106–115.

Eby, L. A., Helmy, O., Holsinger, L. M., & Young, M. K. (2014). Evidence of climate-induced range contractions for bull trout in a Rocky Mountain watershed, U.S.A. *PLoS ONE, 9*, e98812.

Elliott, J. M. (1994). *Quantitative ecology and the brown trout*. New York: Oxford University Press.

Erős, T., Olden, J. D., Schick, R. S., et al. (2012). Characterizing connectivity relationships in freshwaters using patch-based graphs. *Landscape Ecology, 27*, 303–317.

Fausch, K. D., Rieman, B. E., Dunham, J. B., et al. (2009). Invasion versus isolation: Trade-offs in managing native salmonids with barriers to upstream movement. *Conservation Biology, 23*, 859–870.

Ficke, A. D., Myrick, C. A., & Hansen, L. J. (2007). Potential impacts of global climate change on freshwater fisheries. *Reviews in Fish Biology and Fisheries, 17*, 581–613.

Furniss, M. J., Staab, B. P., Hazelhurst, S., et al. (2010). *Water, climate change, and forests: watershed stewardship for a changing climate* (General Technical Report PNW-GTR-844). Portland: U.S. Forest Service, Pacific Northwest Research Station.

Furniss, M. J., Roby, K. B., Cenderelli, D., et al. (2013). *Assessing the vulnerability of watersheds to climate change: results of national forest watershed vulnerability pilot assessments* (General

Technical Report PNW-GTR-884). Portland: U.S. Forest Service, Pacific Northwest Research Station.

Gergely, K. J., & McKerrow, A. (2013). *PAD-US national inventory of protected areas* (U.S. Geological Survey Fact Sheet 2013–3086). Reston: U.S. Geological Survey.

Gresswell, R. E. (2011). Biology, status, and management of the Yellowstone cutthroat trout. *North American Journal of Fisheries Management, 31*, 782–812.

Gresswell, R. E., & Varley, J. D. (1988). Effects of a century of human influence on the cutthroat trout of Yellowstone Lake. *American Fisheries Society Symposium, 4*, 45–52.

Hamlet, A. F., Elsner, M. M., Mauger, G. S., et al. (2013). An overview of the Columbia Basin Climate Change Scenarios Project: Approach, methods, and summary of key results. *Atmosphere-Ocean, 51*, 392–415.

High, B., Meyer, K. A., Schill, D. J., & Mamer, E. R. (2008). Distribution, abundance, and population trends of bull trout in Idaho. *North American Journal of Fisheries Management, 28*, 1687–1701.

Howell, P. J., Dunham, J. B., & Sankovich, P. M. (2010). Relationships between water temperatures and upstream migration, cold water refuge use, and spawning of adult bull trout from the Lostine River, Oregon, USA. *Ecology of Freshwater Fish, 19*, 96–106.

Independent Science Advisory Board (ISAB). (2007). *Climate change impacts on Columbia River Basin fish and wildlife* (ISAB Climate Change Report ISAB 2007-2). Portland: Northwest Power and Conservation Council.

Isaak, D. J., & Rieman, B. E. (2013). Stream isotherm shifts from climate change and implications for distributions of ectothermic organisms. *Global Change Biology, 19*, 742–751.

Isaak, D. J., Luce, C., Rieman, B., et al. (2010). Effects of climate change and recent wildfires on stream temperature and thermal habitat for two salmonids in a mountain river network. *Ecological Applications, 20*, 1350–1371.

Isaak, D. J., Wollrab, S., Horan, D., & Chandler, G. (2012a). Climate change effects on stream and river temperatures across the northwest U.S. from 1980–2009 and implications for salmonid fishes. *Climatic Change, 113*, 499–524.

Isaak, D. J., Muhlfeld, C. C., Todd, A. S., et al. (2012b). The past as prelude to the future for understanding 21st-century climate effects on Rocky Mountain trout. *Fisheries, 37*, 542–556.

Isaak, D. J., Peterson, E., Ver Hoef, J., et al. (2014). Stream isotherm shifts from climate change and implications for distributions of ectothermic organisms. *Global Change Biology, 19*, 742–751.

Isaak, D. J., Young, M. K., Nagel, D., et al. (2015). The Cold-Water Climate Shield: Delineating refugia to preserve salmonid fishes through the 21st century. *Global Change Biology, 21*, 2540–2553.

Kanda, N., Leary, R. F., & Allendorf, F. W. (2002). Evidence of introgressive hybridization between bull trout and brook trout. *Transactions of the American Fisheries Society, 131*, 772–782.

Klemetsen, A. (2010). The charr problem revisited: Exceptional phenotypic plasticity promotes ecological speciation in postglacial lakes. *Freshwater Reviews, 3*, 49–74.

Klemetsen, A., Amundsen, P. A., Dempson, J. B., et al. (2003). Atlantic salmon *Salmo salar* L., brown trout *Salmo trutta* L. and Arctic charr *Salvelinus alpinus* (L.): A review of aspects of their life histories. *Ecology of Freshwater Fish, 12*, 1–59.

Leppi, J. C., DeLuca, T. H., Harrar, S. W., & Running, S. W. (2012). Impacts of climate change on August stream discharge in the Central Rocky Mountains. *Climatic Change, 112*, 997–1014.

Luce, C. H., & Holden, Z. A. (2009). Declining annual streamflow distributions in the Pacific Northwest United States, 1948–2006. *Geophysical Research Letters, 36*, L16401.

Luce, C., Morgan, P., Dwire, K., et al. (2012). *Climate change, forests, fire, water, and fish: Building resilient landscapes, streams, and managers* (General Technical Report RMRS-GTR-290). Fort Collins: U.S. Forest Service, Rocky Mountain Research Station.

Luce, C. H., Abatzoglou, J. T., & Holden, Z. A. (2013). The missing mountain water: Slower westerlies decrease orographic precipitation. *Science, 266*, 776–779.

Luce, C. H., Staab, B. P., Kramer, M. G., et al. (2014). Sensitivity of summer stream temperatures to climate variability in the Pacific Northwest. *Water Resources Research, 50*, 1–16.

Mantua, N. J., & Raymond, C. L. (2014). Climate change, fish, and aquatic habitat in the North Cascade Range. In C. L. Raymond, D. L. Peterson, & R. M. Rochefort (Eds.), *Climate change vulnerability and adaptation in the North Cascades region, Washington* (General Technical Report PNW-GTR-892 (pp. 235–270). Portland: U.S. Forest Service, Pacific Northwest Research Station.

Mantua, N. J., Tohver, I., & Hamlet, A. (2010). Climate change impacts on streamflow extremes and summertime stream temperature and their possible consequences for freshwater salmon habitat in Washington state. *Climatic Change, 102*, 187–223.

Martinez, P. J., Bigelow, P. E., Deleray, M. A., et al. (2009). Western lake trout woes. *Fisheries, 34*, 424–442.

McKelvey, K. S., Young, M. K., Knotek, W. L., et al. (2016a). Sampling large geographic areas for rare species using environmental DNA (eDNA): A study of bull trout *Salvelinus confluentus* occupancy in western Montana. *Journal of Fish Biology, 88*, 1215–1222.

McKelvey, K. S., Young, M. K., Wilcox, T. M., et al. (2016b). Patterns of hybridization among cutthroat trout and rainbow trout in northern Rocky Mountain streams. *Molecular Ecology and Evolution, 6*, 688–706.

Miller, D., Luce, C., & Benda, L. (2003). Time, space, and episodicity of physical disturbance in streams. *Forest Ecology and Management, 178*, 121–140.

Mote, P. W., & Salathé, E. P. (2010). Future climate in the Pacific Northwest. *Climatic Change, 102*, 29–50.

Mote, P. W., Parson, E. A., Hamlet, A. F., et al. (2003). Preparing for climatic change: The water, salmon, and forests of the Pacific Northwest. *Climatic Change, 61*, 45–88.

Northcote, T. G. (1992). Migration and residency in stream salmonids: Some ecological considerations and evolutionary consequences. *Nordic Journal of Freshwater Research., 67*, 5–17.

Peterson, D. P., Rieman, B. E., Horan, D. L., & Young, M. K. (2013a). Patch size but not short-term isolation influences occurrence of westslope cutthroat trout above human-made barriers. *Ecology of Freshwater Fish., 23*, 556–571.

Peterson, D. P., Wenger, S. J., Rieman, B. E., & Isaak, D. J. (2013b). Linking climate change and fish conservation efforts using spatially explicit decision support models. *Fisheries, 38*, 111–125.

Poff, N. L., Brinson, M. M., & Day, J. W. J. (2002). *Aquatic ecosystems and global climate change: Potential impacts on inland freshwater and coastal wetland ecosystems in the United States*. Washington, DC: Pew Center on Global Climate Change.

Rasmussen, J. B., Robinson, M. D., Hontela, A., & Heath, D. D. (2012). Metabolic traits of westslope cutthroat trout, introduced rainbow trout and their hybrids in an ecotonal hybrid zone along an elevation gradient. *Biological Journal of the Linnean Society, 105*, 56–72.

Rieman, B. E., & McIntyre, J. D. (1995). Occurrence of bull trout in naturally fragmented habitat patches of varied size. *Transactions of the American Fisheries Society, 124*, 285–296.

Rieman, B. E., & Isaak, D. J. (2010). Climate change, aquatic ecosystems, and fishes in the Rocky Mountain West: Implications and alternatives for management (General Technical Report GTR-RMRS-250). Fort Collins: U.S. Forest Service, Rocky Mountain Research Station.

Rieman, B. E., Lee, D. C., & Thurow, R. F. (1997). Distribution, status, and likely future trends of bull trout within the Columbia River and Klamath River basins. *North American Journal of Fisheries Management, 17*, 1111–1125.

Rieman, B. E., Isaak, D., Adams, S., et al. (2007). Anticipated climate warming effects on bull trout habitats and populations across the interior Columbia River basin. *Transactions of the American Fisheries Society, 136*, 1552–1565.

Schindler, D. E., Augerot, X., Fleishman, E., et al. (2008). Climate change, ecosystem impacts, and management for Pacific salmon. *Fisheries, 33*, 502–506.

Schrank, A. J., Rahel, F. J., & Johnstone, H. C. (2003). Evaluating laboratory-derived thermal criteria in the field: An example involving Bonneville cutthroat trout. *Transactions of the American Fisheries Society, 132*, 100–109.

Shepard, B. B., Sanborn, B., Ulmer, L., & Lee, D. C. (1997). Status and risk of extinction for westslope cutthroat trout in the upper Missouri River basin, Montana. *North American Journal of Fisheries Management, 17*, 1158–1172.

Shepard, B. B., May, B. E., & Urie, W. (2005). Status and conservation of westslope cutthroat trout within the western United States. *North American Journal of Fisheries Management, 25*, 1426–1440.

Syslo, J. M., Guy, C. S., Bigelow, P. E., et al. (2011). Response of non-native laketrout (*Salvelinus namaycush*) to 15 years of harvest in Yellowstone Lake, Yellowstone National Park. *Canadian Journal of Fisheries and Aquatic Sciences, 68*, 2132–2145.

U.S. Environmental Protection Agency (USEPA). (2014). *Best practices for continuous monitoring of temperature and flow in wadeable streams* (EPA/600/R-13/170F). Washington, DC: Global Change Research Program, National Center for Environmental Assessment.

U.S. Fish and Wildlife Service (USFWS). (2015). *Recovery plan for the coterminous United States population of bull trout (Salvelinus confluentus)*. Portland: U.S. Fish and Wildlife Service.

Ver Hoef, J. M., Peterson, E. E., & Theobald, D. M. (2006). Spatial statistical models that use flow and stream distance. *Environmental and Ecological Statistics, 13*, 449–464.

Wenger, S. J., Luce, C. H., Hamlet, A. F., et al. (2010). Macroscale hydrologic modeling of ecologically relevant flow metrics. *Water Resources Research, 46*, W09513.

Wenger, S. J., Isaak, D. J., Dunham, J. B., et al. (2011a). Role of climate and invasive species in structuring trout distributions in the interior Columbia River Basin, USA. *Canadian Journal of Fisheries and Aquatic Sciences, 68*, 988–1008.

Wenger, S. J., Isaak, D. J., Luce, C. H., et al. (2011b). Flow regime, temperature, and biotic interactions drive differential declines of Rocky Mountain trout species under climate change. *Proceedings of the National Academy of Sciences, USA, 108*, 14175–14180.

Williams, J. E., Neville, H. M., Haak, A. L., et al. (2015). Climate change adaptation and restoration of Western trout streams: Opportunities and strategies. *Fisheries, 40*, 304–317.

Young, M. K., Isaak, D. J., McKelvey, K. S., et al. (2016). Climate, demography, and zoogeography predict introgression thresholds in salmonid hybrid zones in Rocky Mountain streams. *PloS One, 11*, e0163563.

Chapter 5
Effects of Climate Change on Forest Vegetation in the Northern Rockies

Robert E. Keane, Mary Frances Mahalovich, Barry L. Bollenbacher, Mary E. Manning, Rachel A. Loehman, Terrie B. Jain, Lisa M. Holsinger, and Andrew J. Larson

Abstract Increasing air temperature, through its influence on soil moisture, is expected to cause gradual changes in the abundance and distribution of tree, shrub, and grass species throughout the Northern Rockies, with drought tolerant species becoming more competitive. The earliest changes will be at ecotones between life-forms (e.g., upper and lower treelines). Ecological disturbance, including wildfire and insect outbreaks, will be the primary facilitator of vegetation change, and future forest landscapes may be dominated by younger age classes and smaller trees. High-elevation forests will be especially vulnerable if disturbance frequency

---

R.E. Keane (✉) • L.M. Holsinger
U.S. Forest Service, Rocky Mountain Research Station, Missoula, MT, USA
e-mail: rkeane@fs.fed.us; lisamholsinger@fs.fed.us

M.F. Mahalovich
U.S. Forest Service, Northern, Rocky Mountain, Southwestern, and Intermountain Regions, Moscow, ID, USA
e-mail: mmahalovich@fs.fed.us

B.L. Bollenbacher • M.E. Manning
U.S. Forest Service, Northern Region, Missoula, MT, USA
e-mail: bbollenbacher@fs.fed.us; mmanning@fs.fed.us

R.A. Loehman
U.S. Geological Survey, Alaska Science Center, Anchorage, AK, USA
e-mail: rloehman@usgs.gov

T.B. Jain
U.S. Forest Service, Rocky Mountain Research Station, Moscow, ID, USA
e-mail: tjain@fs.fed.us

A.J. Larson
Department of Forest Management, University of Montana, Missoula, MT, USA
e-mail: a.larson@umontana.edu

© Springer International Publishing AG 2018
J.E. Halofsky, D.L. Peterson (eds.), *Climate Change and Rocky Mountain Ecosystems*, Advances in Global Change Research 63,
DOI 10.1007/978-3-319-56928-4_5

increases significantly. Increased abundance and distribution of non-native plant species, as well as the legacy of past land uses, create additional stress for regeneration of native forest species.

Most strategies for conserving native tree, shrub, and grassland systems focus on increasing resilience to chronic low soil moisture, and to more frequent and extensive ecological disturbance. These strategies generally include managing landscapes to reduce the severity and patch size of disturbances, encouraging fire to play a more natural role, and protecting refugia where fire-sensitive species can persist. Increasing species, genetic, and landscape diversity (spatial pattern, structure) is an important "hedge your bets" strategy that will reduce the risk of major forest loss. Adaptation tactics include using silvicultural prescriptions (especially stand density management) and fuel treatments to reduce fuel continuity, reducing populations of nonnative species, potentially using multiple genotypes in reforestation, and revising grazing policies and practices. Rare and disjunct species and communities (e.g., whitebark pine, quaking aspen) require adaptation strategies and tactics focused on encouraging regeneration, preventing damage from disturbance, and establishing refugia.

Keywords Forest productivity • Climate change vulnerabilities • Adaptation strategies and tactics • Conifer forests • Ponderosa pine • Whitebark pine • Lodgepole pine • Grand fir • Douglas-fir • Western white pine • Western red cedar • Green ash • Cottonwood • Limber pine

5.1 Introduction

Climate change will affect vegetation assemblages in the Northern Rockies *directly* through altered vegetation growth, mortality, and regeneration, and *indirectly* through changes in disturbance regimes and interactions with altered ecosystem processes (e.g., hydrology, snow dynamics, nonnative species). Some species may be in danger of decreased abundance, whereas others may expand their range. New vegetation communities may form, and historical vegetation complexes may shift to other locations or become rare.

Here we assess the effects of climate change on forest vegetation, based on species autecology, disturbance regimes, current conditions, and modeling. We focus on important Northern Rockies forest tree species and the vegetation types in Fig. 5.1, inferring the vulnerabilities of each species and vegetation type from information found in the literature. Vulnerability is considered with respect to heterogeneous landscapes, including both vegetation disturbance and land-use history.

5 Effects of Climate Change on Forest Vegetation in the Northern Rockies

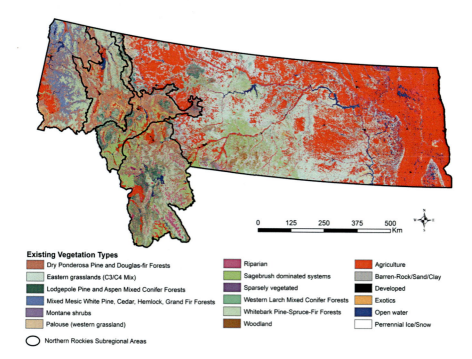

Fig. 5.1 Current vegetation types for the five Northern Rockies subregions. The map was created from the LANDFIRE Existing Vegetation Type map by aggregating National Vegetation Classification Standard vegetation types into a set of vegetation types relevant at coarse spatial scales

5.1.1 Climate Change Assessment Techniques

Past efforts to project the effects of climate change on ecosystem processes have primarily used four techniques (Clark et al. 2001; Schumacher et al. 2006; Joyce et al. 2014). *Expert opinion* involves experts in the fields of climate change, ecology, and vegetation dynamics qualitatively assessing the effects of various climate change scenarios on vegetation. *Field assessment* involves sampling or remote sensing to monitor vegetation change as the climate warms. Although field assessment techniques are the most reliable and useful, they are often intractable because of the large areas and long time periods for which sampling is needed to detect changes. *Statistical analysis* can be used to create empirical models that project climate change response, including projections of habitat, range, or occupational shifts of tree species from climate warming using species distribution models (e.g., Iverson and Prasad 2002). This type of model is inherently flawed, because it relates contemporary species occurrence to current climate, resulting in predictions of *potential* species habitat, not species distribution, and does not include interacting ecological processes (e.g., reproduction, tree growth, competitive interactions, disturbance) (Iverson and McKenzie 2013). *Modeling* to assess climate-mediated vegetation responses is the most effective technique, using projected future climate as inputs into ecological models to simulate climate change effects and interactions (Keane et al. 2004). Models focused on large spatial scales (100–1000 km^2) are best suited for projecting climate change effects, because most ecosystem processes operate and most management decisions are made at large scales (Cushman et al. 2007; McKenzie et al. 2014).

A mechanistic, process-driven simulation approach is needed to emphasize physical drivers of vegetation dynamics directly related to climate, which makes model design complex, with many species characteristics and disturbance factors (Lawler et al. 2006). Ecosystem models that accurately project climate change effects must simulate disturbances, vegetation, climate, and their interactions across multiple spatial scales, but few models simulate ecosystem processes with the mechanistic detail needed to realistically represent important interactions (Keane et al. 2015b; Riggs et al. 2015). A fully mechanistic approach may be difficult for both conceptual and computational reasons, and some simulated processes may always require a stochastic or empirical approach (Falk et al. 2007; McKenzie et al. 2014).

5.1.2 Forest Vegetation Responses to Climate

The effects of climate change on forest vegetation will be driven primarily by altered disturbance regimes, and secondarily through shifts in regeneration, growth, and mortality (Flannigan et al. 2009; Temperli et al. 2013). Trees will respond to reduced water availability, higher temperatures, and changes in growing season in different ways, but because trees are stationary organisms, altered vegetation composition and structure will be the result of changes in plant processes and responses to disturbance.

Several modes of plant function will determine fine-scale response to climate change (Joyce and Birdsey 2000). *Productivity* may increase in some locations because of increasing temperatures and longer growing seasons (especially at higher elevation), but decrease in others where soil moisture decreases (especially at lower elevation). The window of successful *seedling establishment* will change (Ibañez et al. 2007), and increasing drought and high temperatures may narrow the time for effective regeneration in low-elevation forests and widen the window in high-elevation forests. *Tree mortality* can be caused by temperature or moisture stress, as well as late growing-season frosts and high winds (Joyce et al. 2014). *Phenology* may be disrupted in a warmer climate, with some plants experiencing damage or mortality when phenological cues and events are mistimed with new climates (e.g., flowering during dry portions of the growing season). Finally, *genetic limitations* of species or trees may affect their response to climate change (e.g., species restricted to a narrow range of habitat conditions may become maladapted) (Hamrick 2004; St. Clair and Howe 2007) (Table 5.1).

Table 5.1 Comparison of attributes characterizing adaptive strategies for tree species (After Rehfeldt 1994)

Attributes	Adaptive strategy	
	Specialist	Generalist
Factors controlling phenotypic expression of adaptive traits	Genotype	Environment
Mechanisms for accommodating environmental heterogeneity	Genetic variation	Phenotypic plasticity
Range of environments where physiological processes function optimally	Small	Large
Slope of clines for adaptive traits	Steep	Flat
Partitioning of genetic variation in adaptive traits	Mostly among populations	Mostly within populations

Direct effects of temperature on plant growth may increase photosynthesis and respiration. If projected temperatures exceed photosynthetic optima (especially at low elevation), then plant growth might suffer, whereas some trees at high elevation may have photosynthetic gains. Respiration also increases with temperature, so high temperatures coupled with low water availability may result in high respirational losses with few photosynthetic gains (Ryan et al. 1995).

Increased atmospheric CO_2 may increase water-use efficiency (and growth) in some conifer species, potentially compensating for lower water availability. Longer growing seasons and a more variable climate may affect dormancy regulation, bud burst, and early growth (Chmura et al. 2011). Warmer temperatures may reduce growing-season frosts in mountain valleys, thereby allowing cold-susceptible species, such as ponderosa pine (*Pinus ponderosa*), to exist in habitats currently occupied by other species. Snowmelt provides much of the water used by trees in mountain forests, so amount and duration of snowpack will greatly influence regeneration and growth patterns, typically having a negative effect at low elevation and often a positive effect at high elevation (Peterson and Peterson 2001).

Human land-use activities may overwhelm climate change effects in some cases. For example, decades of fire exclusion have resulted in increased tree regeneration and denser canopies in dry forests, coupled with accumulation of fuels (Keane et al. 2002). Because these conditions create competition for water, light, and nutrients, trees in fire-excluded forests are often stressed, making them susceptible to mortality from secondary stressors, such as drought, insect outbreaks, and fire. Most tree species are long lived and genetically diverse, so they can survive wide fluctuations of weather, but interacting drought and modified disturbance regimes will probably play a major role in the future distribution and abundance of forest species (Allen et al. 2010). Most plants have slow migration rates, often depending more on regenerative organs (e.g., sprouting) than seed dispersal. The potential for tree species to migrate may differ among different mountain ranges, depending on local biophysical conditions.

Genetic diversity helps species adapt to changing environments, colonize new areas, and occupy new ecological niches (Ledig and Kitzmiller 1992) (Table 5.1).

Species and populations vulnerable to climate change are typically rare, genetic specialists, species with limited phenotypic plasticity, species or populations with low genetic variation, populations with low dispersal or colonization potential, populations at the trailing edge of climate change, populations at the upper elevation limit of their distribution, and populations threatened by habitat loss, fire, insects, or disease (Spittlehouse and Stewart 2003; St. Clair and Howe 2007). Fragmentation is a critical issue for plant populations because isolation and a small number of individuals can promote inbreeding and loss of genetic diversity (Broadhurst et al. 2008).

5.1.3 Biotic and Abiotic Disturbances

Most changes in vegetation will be facilitated through responses to disturbance or to stress complexes in which multiple factors interact to modify ecosystem structure and function (McKenzie et al. 2009; Iverson and McKenzie 2013; Keane et al. 2015a). Fire exclusion in the Northern Rockies since the 1920s has disrupted annual occurrence, spatial extent, and cumulative area burned by wildfires, resulting in increased surface fuel loads, tree densities, and ladder fuels, especially in low-elevation, dry conifer forests. If drought increases as expected, area burned will increase significantly (McKenzie et al. 2011; Peterson et al. 2014). Reduced snowpack and drier fuels could also make high-elevation forests more susceptible to increasing fire occurrence (Miller et al. 2009).

Insect activity and outbreaks are also affected by climate and will dictate future forest composition and structure. Mountain pine beetle (*Dendroctonus ponderosae*) is an aggressive and economically important insect responsible for high tree mortality across large areas (Logan et al. 2003), and warming temperatures have directly influenced bark beetle-caused tree mortality in much of western North America (Safranyik et al. 2010). Future mortality will depend on spatial distribution of live host trees, heterogeneity of future landscapes, and ability of beetle populations to adapt to changing conditions.

5.2 Climate Change Effects on Tree Species

Climate change effects on trees species in the Northern Rockies were inferred based on autecology, disturbance interactions, and current and historical conditions (Table 5.2). Most information is from published literature, tempered with professional experience. Primary sources of autecological information are Minore (1979), Burns and Honkala (1990), Bollenbacher (2012), and Devine et al. (2012). MC2 model output was used to evaluate climate change effects on important species and vegetation types (Figs. 5.2 and 5.3). The literature is sometimes inconsistent on the response of tree species to climate change, reflecting considerable uncertainty about

5 Effects of Climate Change on Forest Vegetation in the Northern Rockies

Table 5.2 Climate change vulnerability ratings for tree species in the Northern Rockies (including its five subregions); ratings are ordinal, with 1 being the most vulnerable. Species not included in a subregion are indicated by a dash

Species	Northern Rockies	Subregion Western Rockies	Central Rockies	Eastern Rockies	Greater Yellowstone Area	Grassland
Alpine larch	1	2	1	–	–	–
Whitebark pine	2	1	2	1	1	–
Western white pine	3	5	3	–	–	–
Western larch	4	6	4	–	–	–
Douglas-fir	5	8	8	2	2	1
Western redcedar	6	4	5	–	–	–
Western hemlock	7	3	6	–	–	–
Grand fir	8	7	7	–	–	–
Engelmann spruce	9	9	11	3	4	5
Subalpine fir	10	10	12	4	5	6
Lodgepole pine	11	11	10	5	6	7
Mountain hemlock	12	3	9	–	–	–
Cottonwood	13	12	13	6	3	2
Aspen	14	13	14	8	7	3
Limber pine	15	–	15	7	8	4
Ponderosa pine-west	16	14	16	–	–	–
Ponderosa pine-east	17	–	–	9	9	8
Green ash	18	–	–	10	10	9

projections. In addition, the amount of climate change matters. Most climate change studies project minimal changes after moderate warming (B1, B2, A1B, RCP 4.5 scenarios), but major species shifts under extreme emission scenarios (A1, A2, RCP 8.5 scenarios). The time frame used affects the magnitude of response, with most studies projecting much greater changes in vegetation after the mid-twenty-first century.

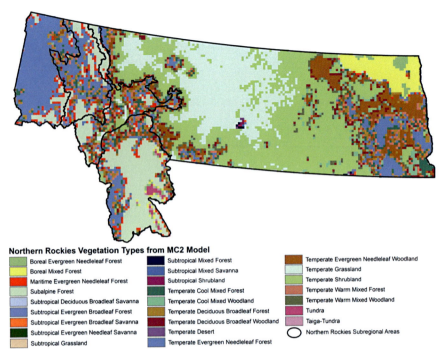

Fig. 5.2 Vegetation types used in the MC2 model within the five Northern Rockies subregions. These vegetation types are represented in the model output in Fig. 5.3

5.2.1 *Ponderosa Pine* (**Pinus ponderosa***)*

Ponderosa pine is a shade-intolerant, drought-adapted species in low-elevation, dry forests of the Northern Rockies (Minore 1979). Ponderosa pine is a "drought avoider" that tolerates dry soil conditions by efficiently closing stomata to avoid water loss and xylem cavitation and stay alive during deep droughts (Sala et al. 2005). Seedlings are highly susceptible to frost damage, and the occurrence of frosts often excludes the pine from low valley settings, especially in frost pockets and cold air drainages (Shearer and Schimidt 1970). As a seral species, ponderosa pine is often associated with Douglas-fir (*Pseudotsuga menziesii*), lodgepole pine (*Pinus contorta* var. *latifolia*), grand fir (*Abies grandis*), and western larch (*Larix occidentalis*). Ponderosa pine is highly resistant to fire, more resistant than nearly all of its competitors (Ryan and Reinhardt 1988), which historically allowed it to maintain dominance over large areas that burned frequently.

Ponderosa pine is expected to tolerate increasing temperatures and droughts with only moderate difficulty. As a "drought avoider," it can close stomata at low soil-water potential, allowing it to persist in low-elevation sites (Stout and Sala 2003). Three studies have projected an expansion of the range of ponderosa pine in a warmer climate (Hansen et al. 2001; Nitschke and Innes 2008; Morales et al. 2015).

5 Effects of Climate Change on Forest Vegetation in the Northern Rockies

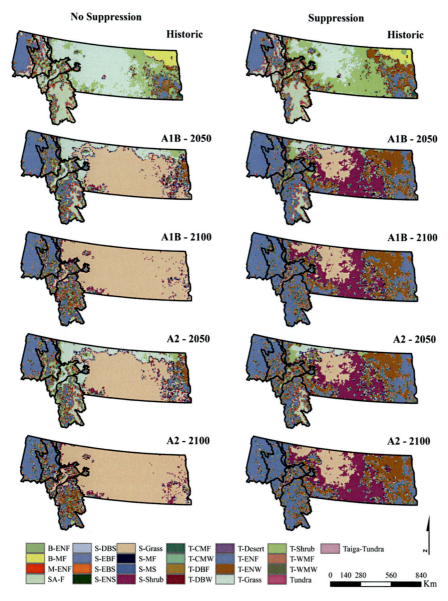

Fig. 5.3 Maps of MC2 vegetation type distributions with and without fire suppression (the two columns) for two emission scenarios (A1B, A2) and three time periods (historical, 2050, 2100) (each row) for five Northern Rockies subregions (outlined in bold on each map). Vegetation types are: *B* boreal, *M* maritime, *SA* subalpine, *S* subtropical, *T* temperate, *ENF* evergreen needleleaf forest, *ENW* evergreen needleleaf woodland, *F* forest, *MF* mixed forest, *MW* mixed woodland, *DBF* deciduous broadleaf forest, *DBW* deciduous broadleaf woodland (not all vegetation types in the legend are represented in the map)

There may also be opportunities for this species to move to higher elevations, based on competitiveness in dry soils (Gray and Hamann 2013). Increases in mountain pine beetle and other insects, advancing competition resulting from fire exclusion, and increases in fire severity and intensity will dictate the future of ponderosa pine in the Northern Rockies. If fires are too frequent, established regeneration will never survive, and mature individuals will not become established. Increasing fire severity and occurrence could also eliminate old trees that provide seed sources for populating future burns.

5.2.2 Douglas-Fir (Pseudotsuga menziesii)

Douglas-fir is a major component of lower elevation and mixed-conifer forests in the Northern Rockies (only Rocky Mountain Douglas-fir [var. *glauca*] is found here]). It is an early-seral species in moist habitats with western larch, western white pine (*Pinus monticola*), grand fir, western redcedar (*Thuja plicata*), and western hemlock (*Tsuga heterophylla*), and is late seral in drier habitats with ponderosa pine, juniper, and quaking aspen (*Populus tremuloides*). Douglas-fir is a "drought tolerator," keeping stomata open to extract soil water at low soil-water potentials, thereby subjecting it to xylem cavitation and potentially death in extreme drought (Stout and Sala 2003; Sala et al. 2005). Western spruce budworm (*Choristoneura occidentalis*), Douglas-fir tussock moth (*Orgyia pseudotsugata*), and Douglas-fir beetle (*Dendroctonus pseudotsugae*) are prominent insects that affect this species. Thick bark, thick foliar buds, and deep main roots confer high resistance to wildfire.

Douglas-fir is expected to have low to moderate vulnerability to climate change. Recent modeling results project no change to significant increases in the range of this species in a warmer climate (Morales et al. 2015), although it is possible that it will decrease in drier portions of its range (Nitschke and Innes 2008). Growth will probably decrease somewhat in a warmer climate, both in the Northern Rockies and the rest of the western United States (Restaino et al. 2016). Increased wildfire, coupled with adverse effects of fire exclusion, could cause mortality in Douglas-fir in stands with high fuel loadings. If fires increase, they may be so frequent that Douglas-fir seedlings cannot become established and become mature trees. Recent surveys show significant increases in Douglas-fir seedling mortality in response to increasing drought and high temperature, which may become more common in the future.

5.2.3 Western Larch (Larix occidentalis)

Western larch grows in moist, cool environments in valley bottoms, benches, and northeast-facing mountain slopes. Larch has low water-use efficiency compared to other conifers in the Northern Rockies, explaining its absence on xeric sites. Early autumn cold snaps affect seedling and sapling survival (Rehfeldt 1995), and drought affects mid- to late-season survival. Larch is a long-lived, shade intolerant, early-seral species, growing fast with tall, open crowns and outcompeting other species (Milner 1992). It is moderately drought tolerant and can survive seasonal drought, but performs poorly when droughts last more than 2 years. Douglas-fir is the most common associate, but many other species can be found with larch. Frequent, low-intensity wildfire historically maintained dominance of larch, which is very tolerant of fire (Ryan and Reinhardt 1988). Extensive logging removed many of the large larch that could have survived fire, and fire exclusion has eliminated the burned mineral soil seedbeds where western larch can regenerate.

Western larch may be susceptible to future changes in climate because of its narrow distribution in the Northern Rockies and its uncertain association with wildfire. Modeling studies suggest that larch may be susceptible to a warmer climate, with potentially large constrictions of its range in some locations (Fins and Steeb 1986; Nitschke and Innes 2008; Aston 2010; Morales et al. 2015). Western larch may migrate to more northerly and higher areas in the Northern Rockies, but not without surviving major fires (Gray and Hamann 2013). Increasing fires may help return western larch to the Northern Rockies landscape, but this would require significant assistance from planting. Continued fire exclusion will probably result in continuing declines of western larch, because increased competition will reduce vigor, making trees more susceptible to insects and pathogens, and fuel loadings will propagate crown fires, causing high larch mortality (Keane et al. 1996).

5.2.4 Western White Pine (Pinus monticola)

Western white pine grows at mid elevations in the Northern Rockies, often in steep topography along moist creek bottoms, lower benches, and north aspects. Intermediate in shade tolerance, it is usually an early-seral species (Minore 1979), attaining dominance in a stand only following wildfire or through silvicultural systems that favor it. Once established, western white pine grows best in full sunlight. Seedlings have low drought tolerance, and seedling mortality in the first growing season is attributed to high surface temperatures and low soil moisture. White pine is tolerant of cold when dormant. Seed germination requires 20–120 days of cold, and occurs following snowmelt, typically on mineral soil (Graham 1990). Mature trees are relatively tolerant of wildfire, especially where they have high, open crowns. The abundance and distribution of white pine is currently restricted because of removal through logging over the past century.

Western white pine may be reasonably well adapted to higher temperature in wetter portions of the Northern Rockies (Loehman et al. 2011). Its fast growth rate and ability to survive fire provide ecological resilience, and it disperses seeds heavily into burned areas, providing an effective means of regeneration with high fire occurrence. However, in much of its range, white pine will be susceptible to declines from interacting effects of fire exclusion, white pine blister rust, and rapid succession to shade tolerant conifer communities. White pine blister rust is a huge stressor, because white pine has not yet developed the genetic capacity to overcome this disease (Fins et al. 2002). Therefore, even if wild fire increases opportunities for regeneration, there may be few residual trees to provide the necessary seed source. Abundance of western white pine is currently low in isolated landscapes, and thus the magnitude of any decline may be large relative to current and past populations Without a comprehensive restoration program, this species may never again be dominant in the Northern Rockies.

5.2.5 *Grand Fir (*Abies grandis*)*

Grand fir is found on a wide variety of sites in the Northern Rockies, including stream bottoms and valley and mountain slopes (Foiles et al. 1990). It is either an early-seral or late-seral species depending on site moisture (Ferguson and Johnson 1996), often found with Douglas-fir, ponderosa pine, western redcedar, western hemlock, and other species. Grand fir is shade tolerant, but is relatively intolerant of drought. It has low frost tolerance but can tolerate seasonally fluctuating water tables. It is susceptible to fire damage in moist creek bottoms, but is more resistant on dry hillsides where roots are deeper and bark is thicker (Ryan and Reinhardt 1988). It is susceptible to heart rot and decay, especially armillaria root rot (*Armillaria* spp.) and annosus root disease (*Heteribasidion annosum*), and is attacked by numerous insects (Foiles et al. 1990). Fire exclusion has greatly increased grand fir on both dry and mesic sites, but increased tree densities have also stressed grand fir trees, contributing to increased fuel loadings, and higher damage and mortality from root rot and insects.

On xeric sites, increased drought and longer growing seasons will exacerbate stress for grand fir, with increased competition and potentially high mortality from insects and disease. Modeling studies have projected major declines in this species by the end of the twenty-first century (Nitschke and Innes 2008; Coops and Waring 2011). However, increased productivity may lead to expanded grand fir populations on sites with moderate moisture (Urban et al. 1993; Aston 2010). On mesic sites where grand fir is seral to western redcedar and western hemlock, longer growing seasons coupled with higher temperatures may increase growth rates and regeneration success. Longer fire seasons and high fuel loadings will potentially rearrange grand fir communities across the Northern Rockies, reducing grand fir dominance at both large and small scales. Although many grand fir forests are stressed from high tree densities, the species will probably tolerate changes in climate and remain on the landscape at levels similar to historical conditions.

5.2.6 Western Redcedar *(Thuja plicata)*

Western redcedar is a component of mesic forests in the Northern Rockies, occupying wet ravines and poorly-drained depressions, often as a riparian species. Shade tolerance is high, and it is often present in all stages of forest succession. It is associated with grand fir, western white pine, western hemlock, western larch, and ponderosa pine, occurring in pure stands only where fire has been excluded for a long time, or where fire has been used to maintain redcedar dominance. It regenerates best on disturbed mineral soil, although scorched soil is not beneficial for regeneration, and seedlings survive best in partial shade. It is not resistant to drought or frost, and can be damaged by freezing temperatures in late spring and early autumn. Western redcedar is not severely affected by most insects and pathogens (Minore 1990), and is moderately fire tolerant when mature.

Western redcedar may retain its current range in a warmer climate, and productivity may increase in cooler, wetter locations (Hamann and Wang 2006; Aston 2010). Although warmer conditions may benefit redcedar in some locations, drier conditions would likely reduce its distribution and productivity in dry to mesic sites, especially if it becomes more susceptible to insect attacks (Woods et al. 2010). Warming could also result in a loss of chilling required for western redcedar growth and reproduction (Nitschke and Innes 2008). The potential effects of disturbance on redcedar are unclear. Fire can maintain redcedar communities if it burns at low severities and kills only seedlings and saplings, but high-severity wildfires can eliminate seed sources. Continued fire exclusion may maintain current western redcedar distributions, but without fuel treatments, crown fires may be sufficiently common to cause extensive mortality. Western redcedar is often associated with ash cap soils, and the potential of redcedar to migrate to non-ash soils under new climates may be limited.

5.2.7 Western Hemlock *(Tsuga heterophylla)*

Western hemlock is found in mild, humid climates and in environments with abundant soil moisture throughout the growing season, typically found in association with western redcedar, grand fir, Douglas-fir, western larch, western white pine, lodgepole pine, and ponderosa pine. Where soils are relatively dry in summer, hemlock is found primarily on north aspects and in moist drainages and other locations where water is available. This species is very shade tolerant and usually considered a late-seral species, although it is often common at all stages of stand development. Hemlock is highly susceptible to drought during the growing season (Baumgartner et al. 1994). It can germinate on a variety of organic and mineral seedbeds, and its seedlings are highly susceptible to frost. Many root and bole pathogens cause significant damage and mortality in western hemlock. It is very susceptible to fire damage because of its shallow roots and thin bark, and its relatively shallow root system makes it susceptible to wind throw. Most stands in the Northern Rockies that contain western hemlock have become denser over the past century, with the hemlock component increasing in the overstory and understory, a condition often leading to reduced vigor.

Increased drought and area burned are expected to reduce abundance and distribution of hemlock, especially in drier locations. Several studies have projected contractions in western hemlock distribution. For example, Hansen et al. (2001) simulated major contractions in western hemlock range, and Shafer et al. (2001) reported that western hemlock may decrease in range because chilling requirements for the seeds will not be met. Keane et al. (1996) simulated losses of western hemlock and redcedar under moderate climate warming in Glacier National Park, mostly as a result of severe fires. Other studies project both a decrease and increase in western hemlock in a warmer climate (Urban et al. 1993; Cumming and Burton 1996; Hamann and Wang 2006). It is possible that western hemlock will maintain most of its current range in the future, although it may not have the diversity in growth habit to allow it to expand its range into higher-elevation sites as temperatures warm.

5.2.8 Lodgepole Pine *(Pinus contorta var. latifolia)*

Lodgepole pine has the widest range of environmental tolerance of any conifer in North America (Lotan and Critchfield 1990) and is found in a broad range of soils and local climatic conditions in the Northern Rockies. Shade intolerant and relatively tolerant of both drought and cold temperatures, this species grows in nearly pure stands as well as in association with several other conifer species. The presence of cone serotiny in most populations allows lodgepole pine to reproduce prolifically following wildfire, and seedlings can survive diverse microsite and soil conditions, although drought is a common cause of mortality in first-year seedlings. Fire plays a critical role in lodgepole pine forest succession, and many current forests originated from stand-replacement fires. Mature trees have moderate tolerance to fire and can survive light burns. Mountain pine beetle also plays a significant role in the dynamics of lodgepole pine ecosystems, as evidenced by recent large outbreaks in the Northern Rockies and much of western North America, which have resulted from a combination of increased temperature and an abundance of low-vigor stands (Carroll et al. 2003).

Longer droughts and warmer temperature in lower-elevation sites may reduce lodgepole pine growth and regeneration, with a possible transition to other tree species (Chhin et al. 2008; Nigh 2014). The results of different modeling studies are equivocal about the future distribution of this species in a warmer climate, but given that lodgepole pine is a generalist capable of regenerating and growing in a wide range of environments, it is likely that the decline of lodgepole pine from drier sites will occur only under extreme warming scenarios over long time periods. In the subalpine zone where seasonal drought is not a problem, moderate warming may increase lodgepole pine productivity (Johnstone and Chapin 2003; Wang et al. 2006; Aston 2010) and possibly distribution. The ultimate fate of lodgepole pine will depend on frequency and extent of wildfire (Smithwick et al. 2009). Populations with serotiny can be expected to respond well to future fire, but very frequent fire could eliminate younger stands. In addition, mountain pine beetle outbreaks will probably have a major influence on lodgepole pine abundance and distribution in a warmer climate (Creeden et al. 2014). As with fire, the frequency and extent of beetle-caused mortality will dictate future stand conditions (Logan and Powell 2001). In summary, lodgepole pine distribution may both expand and contract depending on location, but the species is expected to persist in the Northern Rockies as long as fire remains on the landscape.

5.2.9 *Limber Pine* (Pinus flexilis)

Limber pine is a shade-intolerant, early-seral species in the Northern Rockies (Steele 1990). Limber pine has difficulty competing with other species on more productive mesic sites and is often succeeded by Douglas-fir and subalpine fir. Reproduction is often very low for this slow-growing, long-lived species whose seeds are dispersed by rodents and Clark's nutcracker (Lanner 1980). Limber pine is tolerant of drought and can become established and grow in arid environments. The broad niche occupied by limber pine indicates this species has a generalist adaptive strategy. However, it is experiencing significant damage from white pine blister rust and mountain pine beetle in some locations (Taylor and Sturdevant 1998; Jackson et al. 2010).

Warming temperatures and decreasing snowpack will result in increased growth in many limber pine communities (Aston 2010). Increases in vigor are usually accompanied by larger cone crops, higher seed viability, more seeds per cone, wider seed dispersal, and greater resistance to disease. Warm temperature could cause drier soils, especially for seed germination and seedling growth. Disturbance interactions will affect limber pine dynamics in a warmer climate. Increased wildfire may limit its encroachment into grasslands in areas where grazing is low. Warmer, drier conditions may also reduce blister rust infection by disrupting the pathogen life cycle, especially during the late summer when infection occurs.

5.2.10 *Subalpine Fir* (Abies lasiocarpa)

Subalpine fir occupies lower valleys to the upper subalpine zone in the Northern Rockies, often associated with grand fir, western larch, Douglas-fir, western redcedar, and western white pine at lower elevations (Pfister et al. 1977) and with lodgepole pine, Engelmann spruce, whitebark pine, alpine larch, and mountain hemlock at higher elevations (Arno 2001). Fir tolerates shade, but cannot tolerate prolonged drought, especially in the seedling stage. Reproduction tends to occur in pulses relative to periodic seed crops and the occurrence of favorable weather for germination and establishment (Alexander et al. 1990). Fir is highly susceptible to fire damage because of thin bark, dense foliage, and shallow roots (Ryan and Reinhardt 1988), and even low-severity fires can cause high mortality. Several insects and pathogenic fungi damage this species, especially in older, low-vigor stands. Abundance of subalpine fir has increased in some Northern Rockies landscapes (Keane et al. 1994), increasing stress from competitive interactions and causing at least some mortality during dry periods.

With a diverse range throughout the Northern Rockies, subalpine fir could expand its range into the treeline, become more productive in colder portions of its current range, and decline in growth and extent in warmer, drier portions of its current range. Model output ranges from large losses of subalpine fir (Hamann and Wang 2006) to minimal change in its distribution (Bell et al. 2014). Most paleo-reconstructions in the Holocene show that subalpine fir was dominant during cold periods and declined during warm periods (Whitlock 1993, 2004; Brunelle et al. 2005). The future of subalpine fir will depend on the degree of warming and frequency and extent of disturbance, especially wildfire. Increased fire would reduce subalpine fir dominance faster and more extensively than direct climate effects. This species may shift across the high mountain landscape, with gains balancing losses (caused directly by changes in climate). However, future increases in fire, disease, and insects may limit its abundance. Because fir is an aggressive competitor, gains through advanced succession in the upper subalpine zone may balance or exceed losses from fire, drought, and pathogens in the lower subalpine zone.

5.2.11 Engelmann Spruce (Picea engelmannii*)*

Engelmann spruce is a major component of high-elevation forests in the Northern Rockies, and although it commonly occurs with subalpine fir, it is also associated with many other conifer species. Spruce is shade tolerant and cold tolerant, but is intolerant of low soil moisture and prolonged drought (Alexander and Shepperd 1990). Seedlings are very intolerant of high temperatures and low soil moisture. Spruce is very susceptible to fire injury and mortality, although some mature trees can survive fire (Bigler et al. 2005), thus providing a post-fire seed source. Spruce beetle (*Dendroctonus rufipennis*) and western spruce budworm are serious stressors, usually attacking older, low-vigor trees. Logging and fire have reduced spruce in some lower-elevation areas in the Northern Rockies.

In a warmer climate, some losses of Engelmann spruce may occur in drier portions of its range, especially in seasonally moist sites. Mortality events in Engelmann spruce over the last 20 years have been attributed to prolonged drought, presumably related to changing climate (Liang et al. 2015), and warm, dry weather has been associated with periods of low growth (Alberto et al. 2013). Most modeling output suggests that spruce will decrease in distribution while moving up in elevation. Spruce may become established in high-elevation locations where snow precluded conifer regeneration historically (Schauer et al. 1998), particularly because it has the genetic capacity to adapt to large swings in climate (Jump and Peñuelas 2005). With good seed dispersal and tall stature, spruce is able to establish in previously non-forested areas. Paleoclimatic studies indicate that spruce regeneration was highest during the warmest (low snow) periods of the past several centuries. Because spruce is not resistant to wildfire, major declines could occur if projected increases in fire reach forests where spruce is dominant. Spruce beetle will be an ongoing stressor.

5.2.12 Mountain Hemlock (Tsuga mertensiana)

Mountain hemlock is found in cold, snowy upper subalpine sites where it grows slowly and can live to be more than 800 years old. It is commonly associated with subalpine fir and lodgepole pine, often restricted to north slopes. Hemlock is very shade tolerant, competes well with other species, and is usually considered a late-seral species (Minore 1979; Means 1990). It is not fire resistant, because although it has thick bark when mature, it retains low branches and has shallow roots (Dickman and Cook 1989). Hemlock is susceptible to laminated root rot (*Phellius weirii*) which can rapidly kill large groups of trees (Means 1990). Fire exclusion has probably facilitated an increase in hemlock in some locations in the Northern Rockies.

In a warmer climate, mountain hemlock forests are expected to increase in productivity at the highest elevations, but could experience some drought stress at lower elevations (Peterson and Peterson 2001). This potential for increasing productivity at high elevation may buffer mountain hemlock from a warmer climate for many decades. Higher temperatures and less snowpack would also facilitate increased regeneration (Woodward et al. 1995). The potential effects of fire are a big uncertainty. Mountain hemlock has a limited range in the Northern Rockies, so if warming and drying facilitate increased spread of fire into higher subalpine habitats, then hemlock could be threatened.

5.2.13 Alpine Larch (Larix lyallii)

Alpine larch is a deciduous conifer that occupies the highest treeline environments in the Northern Rockies, including the Bitterroot, Anaconda-Pintler, Whitefish, and Cabinet Ranges of western Montana. Larch grows in cold, snowy, and generally moist climates, often in pure stands but also associated with whitebark pine, subalpine fir, and Engelmann spruce. Because this species relies on subsurface water in summer, it has very low drought tolerance (Arno 1990). This shade intolerant conifer has a high capacity to survive wind, ice, and desiccation damage during winter when needles are off the trees, and seedlings are very cold tolerant. Fuel loadings are typically low in the subalpine zone, so large fires are infrequent, killing or injuring larch when they do occur. Alpine larch populations have been relatively constant historically, although the species may be increasing in ribbon forest glades and high-elevation areas where snowpack has been low during the past 20 years.

Alpine larch is expected to be susceptible to climatic shifts that result in increasing drought and fire. As a shade- and drought-intolerant species, alpine larch is not expected to be competitive in increasingly drier soils (Arno and Habeck 1972). Low water availability would be especially damaging to larch in the southern portions of its range, and other subalpine species may be more competitive under stressful conditions. Larch is not well-adapted to survive wildfire (Arno 1990), and if fires become more frequent, this species would experience considerable mortality. However, alpine larch is a prolific seeder and may be able to take advantage of new seedbeds at treeline that were historically covered with snow most of the year. In addition, like other subalpine species, larch may grow faster in a warmer climate. Where alpine larch is able to genetically intergrade with western larch, hybrids may be more tolerant of drought and competition (Carlson et al. 1990).

5.2.14 Whitebark Pine (**Pinus albicaulis***)*

Whitebark pine is an important component of upper subalpine forests in the Northern Rockies—a keystone species that supports high community diversity (Tomback et al. 2001). It is associated with subalpine fir, Engelmann spruce, and mountain hemlock. This species is slow growing, long lived and moderately shade tolerant (Minore 1979), surviving extended drought, strong winds, thunderstorms, and blizzards (Callaway et al. 1998). It occurs as krummholz and small tree islands at exposed treeline sites. Whitebark pine regeneration benefits from the Clark's nutcracker burying thousands of pine seeds in "seed caches" across diverse forest terrain (Keane et al. 2012). Whitebark pine fire regimes are complex and variable in space and time, creating diversity in age, stand structure, and habitat characteristics (Keane et al. 1994). Mountain pine beetle is the most damaging insect in mature stands, often spreading upward from lodgepole pine forests. A severe epidemic caused high mortality in whitebark pine in the Northern Rockies between 1909 and 1940, and mortality has been high in the Greater Yellowstone Area in recent years. White pine blister rust has killed large numbers of whitebark pine in the Northern Rockies, with mortality exceeding 80% in some locations (Keane et al. 2012). Efforts to propagate rust-resistant pines have led to recent plantings of resistant nursery stock in areas that have burned or were declining.

The fate of whitebark pine in a warmer climate will largely depend on local changes in disturbance regimes and their interactions (Keane et al. 2015a). Although this species has a limited range, it was able to persist through many climatic cycles in the past (Whitlock and Bartlein 1993; Whitlock et al. 2003). The predominant stress of blister rust, which precludes regeneration in burned areas, is the greatest cause for concern individually and in combination with climate change. Recent mortality from blister rust and mountain pine beetle have been widespread in the

Northern Rockies (Keane and Parsons 2010). A warmer climate is expected to exacerbate this decline because (1) pine is confined to upper subalpine environments (2) its populations are low, and (3) its regeneration is limited. The only realistic pathway for maintaining viable populations of whitebark pine in the future is for rust-resistant individuals to survive, supplemented by restoration efforts, in order to propagate stands that are resilient enough to survive to reproduction age.

5.2.15 Quaking Aspen **(Populus tremuloides)**

The most widely distributed tree species in North America, quaking aspen is abundant in the mountains of western and southwestern Montana and northern Idaho. Aspen is a short-lived, shade-intolerant, disturbance-maintained seral species. It sprouts aggressively following any disturbance (usually fire) that kills most of the live stems, thus stimulating vegetative propagation (Bartos 1978). Parent trees produce stems, resulting in a clone of genetically identical stems, a reproductive strategy that allows aspen to establish quickly on disturbed sites and out-compete conifers (Mitton and Grant 1996; Romme et al. 1997). Since around 1970, aspen has been in a period of general decline that is thought to be the result of wildfire exclusion, which has allowed plant succession to proceed toward conditions that ordinarily exclude aspen (Frey et al. 2004).

Quaking aspen may experience both gains and losses in a warmer climate, depending on local site conditions. Aspen on warmer, drier sites could experience high mortality because of increasing water deficit (Ireland et al. 2014). Sudden aspen decline has been associated with prolonged drought, particularly in aspen stands that are on the edge of its distribution (Frey et al. 2004). Stress complexes with extreme weather (drought, freeze-thaw events), insect defoliation, and pathogens may be particularly damaging (Brandt et al. 2003; Marchetti et al. 2011), and areas with high ungulate herbivory may have little regeneration. Increased fire frequency, particularly on moist sites, will favor aspen regeneration in the future by removing conifers. However, if future fires are severe, they may kill the shallow root systems and eliminate aspen. Areas with mountain pine beetle-caused conifer mortality may release aspen regeneration once the conifer canopy is thinned or removed, assuming sufficient soil moisture is available.

5.2.16 Cottonwood **(Populus spp.)**

Black cottonwood (*Populus trichocarpa*) and narrowleaf cottonwood (*P. angustifolia*) grow primarily on seasonally wet to moist, open-canopy sites, typically in riparian areas in the western portion of the Northern Rockies. Plains cottonwood (*P. deltoides*) occupies similar habitat in eastern Montana and the Dakotas. Cottonwood is very shade intolerant, and shade-tolerant conifers can encroach and become dominant in upland cottonwood forests (e.g., river and stream terraces). It is also drought intolerant, and requires reliable access to the water table during the growing season (Rood et al. 2003). Plains cottonwood is probably more resilient to drought than the other species. High streamflows and deposition of alluvial sediments are required for seedling establishment, and all cottonwood species are prolific producers of windborne seed. Cottonwood is mildly fire tolerant owing to thick bark and high branches, but is a weak sprouter (Brown 1996). Although several insects attack cottonwood, tent caterpillars (*Malacosoma* spp.) are the only important foliar feeders. Many fungal species can cause decay. Black cottonwood is less common today than it was historically.

In a warmer climate, lower snowpacks will alter streamflows, which may in turn affect germination and establishment of young cottonwoods (Whited et al. 2007). Any alteration of hydrologic flow regime will affect both floodplain interaction and available water (Beschta and Ripple 2005). Effects on regeneration could be positive or negative, depending on the frequency and magnitude of flooding and alluvial deposition. Higher human demands for water could also affect water supplies in riparian areas. Some streamflow-floodplain interactions could result in a conversion of streamside vegetation from cottonwood to upland species (Beschta and Ripple 2005). Plains cottonwood, which currently grows in more arid locations, may be more persistent in a warmer climate because it tends to grow in finer-textured soils that retain sufficient water.

5.2.17 Green Ash (**Fraxinus pennsylvanica**)

Green ash is restricted to the northern Great Plains, which is the northwestern edge of its distribution. Typically found in riparian areas and floodplains, it is well-adapted to climatic extremes and has been widely planted in the Plains states and Canada. In the northern Great Plains, it grows best on moist, well-drained alluvial soils, but is found in other topographic positions where some subsurface water is available. Green ash is moderately shade tolerant in woody draws and is considered an early-seral species. It can propagate vegetatively through stump sprouting, which provides resilience to both mechanical damage from flooding and to occasional wildfire (Lesica 2009). The species is relatively drought tolerant, although prolonged drought inhibits regeneration. Green ash stems are easily killed by fire, but stumps of most size classes of green ash sprout readily after fire (Lesica 2009). Some green ash communities on the western fringe of the northern Great Plains may be declining compared to historical levels (Lesica 2001).

Green ash has a broad ecological amplitude and can survive low soil moisture, but grows optimally on moist sites. In a warmer climate, marginal sites may become less favorable for regeneration and survival of young trees. Higher temperatures may increase ash growth, as long as sufficient water is available. Increased fire frequency would reduce reproduction by seedlings although most mature trees would persist through sprouting. Browsing pressure on green ash communities may increase with increased drought, as upland grasses and forbs desiccate and senesce earlier, or are replaced by invasive, less palatable species. The biggest threat to ash may be the non-native emerald ash borer (Agrilus planipennis), which is spreading westward across North America and may reach the Northern Rockies within the next decade.

5.3 Effects of Climate Change on Broader Vegetation Patterns

The assessment of species vulnerabilities discussed above can be aggregated to assess the vulnerability of broader vegetation assemblages to climate change (Figs. 5.2 and 5.3). Understanding climate change response at this higher level is critical because vegetation assemblages (groups of species) are the focus of most forest management and restoration. The assessment below focuses on dominant vegetation types used by the U.S. Forest Service (USFS) Northern Region.

Dry ponderosa pine/Douglas-fir forests, already located in dry regions, are expected to have significant effects in a warmer climate. Some or all of the tree species may expand into the mixed mesic forest type (next section), especially on south slopes, as drought increases. Fire exclusion has resulted in forest densification and accumulation of surface fuels that will likely support high-severity fires in future decades (Keane et al. 2002). With increasing fire, much of this vegetation type could see losses of Douglas-fir and increases in ponderosa pine. Dry Douglas-fir communities that are currently too cool to support ponderosa pine may support more ponderosa pine in the future.

Western larch mixed conifer forests, found in northern Idaho and northwestern Montana, have been greatly altered from their historical structure. Fire exclusion, coupled with climate change, will probably continue to reduce western larch and increase the more shade-tolerant Douglas-fir, grand fir, and subalpine fir in some areas. Continued fire exclusion will result in further accumulation of fuels, increasing risk of high-severity fire. Western larch is not susceptible to the many insects and diseases common in associated tree species, and is very fire tolerant. However, this species dominates cooler, wetter topographic positions, and a warming climate may constrain the distribution of larch to only north aspects and other microhabitats capable of retaining sufficient water during the growing season (Rehfeldt and Jaquish 2010). This vegetation type will be susceptible to climate-induced increases in area burned by wildfire for the foreseeable future, unless stand structure and landscape pattern can be managed to improve resilience to higher temperatures and higher levels of disturbance.

Mixed mesic western white pine, cedar, hemlock, grand fir forests provide an important context for assessing the effects of climate change. As moist forests experience climate change, competition among species will be dynamic at both small and large spatial scales. A logical approach is to identify specific landscape components that may respond in a coherent manner—north slopes vs. south slopes, side slopes vs. valley bottoms, etc.—and how environmental niches change over time and space. Species that require high soil moisture (hemlock, redcedar) may over time become less common, with drought tolerant species such as Douglas-fir and ponderosa pine becoming more common (Graham 1990). The frequency and magnitude of disturbance, especially fire, will determine composition and structure in these forests. In the short term, we expect more and larger crown fires. In the long term, we expect that frequent fire will favor fire-resistant tree species, maintain a more open forest structure, and maintain younger age classes.

Lodgepole pine mixed subalpine forests, located at high elevations along and east of the Continental Divide, are expected to have relatively low vulnerability to climate change, depending on if and how disturbance is altered. Productivity of subalpine species may increase in a warmer climate, provided that sufficient water is available during the growing season. Species composition may shift slightly, but lodgepole pine, subalpine fir, and quaking aspen will probably still dominate high mountain landscapes for the foreseeable future. Mountain pine beetle outbreaks may become increasingly chronic in a warmer climate, at least in the next few decades, and may continue to reduce the dominance of old lodgepole pine stands. If wildfire is not excluded from this forest type, composition and structure will generally be more resilient to climate change.

Whitebark pine mixed upper subalpine forests will respond more to whitebark pine mortality from white pine blister rust than to climate change, with significant changes in forest composition and structure. Over the last 40 years, whitebark pine has become a minor component of this forest type in many parts of the western Northern Rockies because of blister rust, allowing subalpine fir to become dominant. Recent fires in the upper subalpine zone have reset succession to early-seral stages of shrub and herbaceous communities, but whitebark pine regeneration levels are low because of low population levels (Retzlaff et al. 2016), keeping burned areas in the shrub/herb stage for long periods. We expect that species dominance will continue to shift to subalpine fir, Engelmann spruce, and lodgepole pine. Many Northern Rockies wilderness areas have lands above the elevations at which this forest type occurs, so there are potential areas for range expansion. Some wildfire is needed to create conditions in which whitebark pine can become established and grow to maturity, but if fires are too severe, they will kill the pines needed to provide seeds for regeneration. Planting with rust-resistant trees will be needed to ensure the persistence of whitebark pine in this forest type.

5.4 Natural Resource Issues and Management

5.4.1 Landscape Heterogeneity

High landscape heterogeneity creates diverse biological structure and composition that are considered more resilient and resistant to disturbances (Cohn et al. 2015). For example, the effects of mountain pine beetle outbreaks are less severe in landscapes with diverse age structures of host tree species (Schoettle and Sniezko 2007).

Heterogeneous landscapes also promote population stability, because fluctuations in plant and animal population are less when landscape structure is diverse (Turner et al. 1993). Heterogeneous landscapes may also have more corridors, buffers, and refugia for wildlife and plant migration.

During the past 100 years, land management practices have altered the temporal and spatial characteristics of Northern Rockies landscapes. Timber management has modified patch shape and structure at lower elevation, and fire exclusion has changed patch size and diversity. Fire exclusion has in many cases created landscapes with large contiguous patches of old, dense stands with high fuel accumulations (Keane et al. 2002), although some areas with frequent disturbance are also homogeneous compared to pre-settlement forests. Many forests currently in late-seral conditions have low vigor and high fuel accumulations, making them susceptible to insects and disease and to the risk of severe wildfire.

Many lower-elevation forests in the Northern Rockies have less ability to buffer potential climate change effects because of high stand densities and dominance by shade-tolerant species. However, many higher-elevation forests, especially in the subalpine zone, have species composition and structure similar to what they were historically. Although recent wildfires, restoration activities (thinning, prescribed burning), and timber harvest have helped return some heterogeneity, most landscapes are outside their historical range and variability (HRV) in landscape structure (Box 5.1). This is a significant impediment to improving resilience to the stresses expected from climate change.

Landscape heterogeneity may increase if climate-mediated changes in disturbance regimes increase (Funk and Saunders 2014). Wildfire area burned and mountain pine beetle outbreaks have increased over the past 20 years, in some cases replacing late-seral forests with younger forests with more diverse structure. Continued increases in disturbances (Marlon et al. 2009; Bentz et al. 2010) might balance any loss of biodiversity with gains in landscape heterogeneity (Kappelle et al. 1999).

Large wildfires that will inevitably burn Northern Rockies landscapes may create large patches of homogeneous post-burn conditions (Flannigan et al. 2005, 2009), or may result in semi-permanent shrublands and grasslands in areas too dry for rapid conifer establishment (Fulé et al. 2004). However, post-fire heterogeneity varies considerably across large landscapes (Keane et al. 2008), and the extent of wildfires will almost certainly overwhelm any patterns created by land management.

Using HRV of landscape characteristics is a straightforward approach for making most forests more resilient to climate change (Keane et al. 2009; Keane 2013) (Box 5.1). Although HRV may not represent future conditions, it represents landscape conditions that have proven durable for centuries to millennia (Landres et al. 1999). HRV can be initially used as a reference for restoration (Keane et al. 2015a), then ecological models can be used to project future range of variability for a particular forest location (Keane 2012).

Box 5.1 Using Historical Range and Variability to Assess and Adapt to Climate Change

To effectively implement ecosystem-based management, land managers often find it necessary to identify a reference or benchmark to represent the conditions that describe fully functional ecosystems. Contemporary conditions can be evaluated against this reference to determine status, trend, and magnitude of change, and to design treatments that provide ecosystem services while returning declining ecosystems to a more sustainable condition. Reference conditions are assumed to represent the dynamic character of ecosystems and landscapes, varying across time and space.

The concept of **historical range and variability (HRV)** was introduced in the 1990s to describe past spatial and temporal variability of ecosystems, thus providing a foundation for planning and management. HRV has sometimes been equated with "target" conditions, although targets can be subjective and somewhat arbitrary, representing only one possible situation from a range of potential conditions.

HRV represents a historical envelope of possible ecosystem conditions—burned area, vegetation cover type area, patch size distribution—that can provide a time series of reference conditions. This assumes that (1) ecosystems are dynamic, and their responses to changing processes are represented by past variability; (2) ecosystems are complex and have a range of conditions within which they are self-sustaining, and beyond this range they transition to disequilibrium; (3) historical conditions can serve as a proxy for ecosystem health; (4) time and space domains that define HRV are sufficient to quantify observed variation; and (5) ecological characteristics assessed for ecosystems or landscapes match the management objective.

The use of HRV has been challenged because a warmer climate may permanently alter the environment of ecosystems beyond what was observed under historical conditions, particularly altered disturbance processes, shifts in plant species distribution, and hydrologic dynamics. However, a critical evaluation of possible alternatives suggests that HRV is still a viable approach in the near term, because it has relatively lower uncertainty than methods that predict future ranges of variability.

An alternative to HRV is projecting **future range and variability (FRV)** for landscapes under changing climates, using empirical and mechanistic models. However, the range of projections for future climate from global climate models may be greater than the variability of climate over the past three centuries. This uncertainty increases when projected responses to climate change through technological advances, behavioral adaptations, and population growth are included. Moreover, variability of climate extremes, which will drive most ecosystem response to climate-mediated disturbance and plant dynamics, is difficult to project. Uncertainty will increase as climate projections are extrapolated to the finer scales and longer time periods needed to quantify FRV for landscapes.

> Given these cumulative uncertainties, time series of HRV may have lower uncertainty than simulated projections of future conditions, especially because large variations in past climates are already captured in the time series. It may be prudent to wait until simulation technology has improved enough to create credible FRV landscape pattern and composition, a process that may require decades. In the meantime, attaining HRV would be a significant improvement in functionality of most ecosystems in the Northern Rockies, and would be unlikely to result in negative outcomes from a management perspective. As with any approach to reference conditions, HRV is useful as a guide, not a target, for restoration and other management activities.

5.4.2 Timber Production

Approximately 22,000 km^2 of forested lands are currently managed for timber in the USFS Northern Region, reflecting a large decrease over the past 30 years. Species composition of timber harvests has fluctuated, with harvest following tree mortality caused by disturbance agents such as mountain pine beetle (lodgepole pine), spruce beetle (Engelmann spruce), white pine blister rust (western white pine), root disease (Douglas-fir, grand fir), and wildfire (several species). The current amount of land in each of the major species in lands suitable for timber production is ponderosa pine (6%), dry Douglas-fir (13%), lodgepole pine (27%), western larch (6%), subalpine fir and Engelmann spruce (12%), and mixed western white pine, grand fir, western hemlock, moist Douglas-fir and western redcedar (35%).

Recent harvests in mixed mesic forest are removing grand fir, Douglas-fir and western hemlock, and replanting western white pine, western larch and ponderosa pine. Other harvests involve removal of lodgepole pine and replanting of western larch. Thinning in ponderosa pine and dry Douglas-fir forests is also common. In eastern Montana and the Greater Yellowstone Area, harvesting has focused on beetle-killed lodgepole pine and ponderosa pine. Commercial and restoration thinning in ponderosa pine and dry Douglas-fir is also common.

A large amount of forested lands suitable for timber harvest is in mesic montane and subalpine forests, where productivity may increase in a warmer climate (Aston 2010), potentially leading to higher timber value. However, these forests could also become denser, less productive for timber, and more susceptible to insects and disease, especially in the absence of fire or active management (Joyce et al. 2008). In the future, harvesting timber from mature stands might be a race against losses from disturbance agents (Kirilenko and Sedjo 2007). Simply having more fire and smoke across the landscape in the future will limit access and opportunities for timber harvest.

It is essential that ecological principles be used to design harvest treatments to ensure that future forests are resilient to a warmer climate while continuing to provide a sustainable source of wood. Multiple resources and ecosystem services will need to be considered as well, including residual fuel loadings, soil fertility, water quality, wildlife habitat, and fisheries habitat. Some practices that confer resilience for a particular resource may conflict with other objectives, requiring interdisciplinary planning to find optimal solutions.

5.4.3 Carbon Sequestration

Storage of carbon in (living and dead) biomass and in soils to reduce and defer carbon emissions into the atmosphere is an increasingly important consideration in forest management. Forests in the United States currently offset about 15% of annual U.S. carbon emissions. Size and persistence of forest carbon sinks depend on land management, vegetation composition and structure, and disturbance processes. Although long intervals between disturbance events allow carbon to accumulate for long periods of time, probability of disturbance increases with time (Loehman et al. 2014). Disturbance-prone forests will eventually emit stored carbon, regardless of management intervention, and net carbon balance is near zero over long time periods and large landscapes—unless changes in ecosystem structure and function occur.

This means that (1) disturbance-prone systems cannot be managed to increase stored carbon over historical amounts without limiting disturbance, and (2) shifts in vegetation abundance and distribution will alter spatial patterns of carbon storage. Therefore, expectations for carbon storage need to be developed in the context of climate change effects on vegetation, disturbance, and their interactions.

In general, expected increases in wildfire and other disturbances in the Northern Rockies will make it extremely difficult to maintain forest carbon storage at or above historical levels. Potential for future carbon storage can be assessed as follows:

- Is it reasonable to expect the system to accumulate carbon over historical levels, if the frequency, severity, and magnitude of disturbance events increases?
- What are appropriate temporal and spatial scales over which to measure carbon storage?
- Can potential future disturbance events be managed? Will it be possible to suppress or exclude wildfires, and at what economic or ecological costs?
- Can the effects of additional stressors (drought, invasive species, etc.) be mitigated to help maintain existing vegetation?
- Are future climatic conditions conducive to persistence of forests, or will conditions be inhospitable for current species?
- Do carbon accounting methods assess benefits of natural disturbance processes in carbon-equivalent units that can be weighed against carbon losses?

These are challenging questions that need to be informed by empirical data, ecosystem modeling, and future monitoring to incrementally improve our understanding of ecological drivers and responses to disturbance (Loehman et al. 2014).

Monitoring data can also be used to calibrate, validate, and provide input to models. Models can be used to simulate emergent environmental patterns, compare effects of potential treatments, and identify vulnerable landscapes or ecosystem components.

5.5 Adapting Forest Vegetation and Management to Climate Change

Adaptation to climate change can be defined as initiatives and measures to reduce the vulnerability of natural and human systems against actual or expected climate change effects (IPCC 2007). Most land managers have the tools, knowledge, and resources to begin to address climate change, which requires considering new issues, spatial scales, timing, and prioritization of efforts beyond a steady-state worldview (Swanston and Janowiak 2012).

Risk management is a key component of adaptation, prioritizing actions based on the magnitude and likelihood of climate change effects on resource vulnerability. *Adaptive management* provides a decision-making framework that maintains flexibility and incorporates new knowledge and experience over time. *No-regrets actions* focus on low-risk implementation of projects that could produce multiple benefits, regardless of climate change implications (e.g., removal of invasive species). *Triage* is sometimes needed in situations where vulnerability is high and immediate action is needed (e.g., a species facing extirpation). *Accomplishing multiple objectives* is often possible where an adaptation action also provides benefits for other resource objectives (e.g., riparian restoration, fuel treatments). *Addressing uncertainty* is a necessary component for adaptation, as for most resource planning, guiding the scope and timing of implementation.

A workshop process was used to identify adaptation options for all resources in the Northern Rockies, including vegetation. Teams of resource specialists and scientists reviewed climate change scenarios and a recent scientific assessment of the effects of climate change on vegetation. In response, they developed adaptation strategies (overarching, general) and adaptation tactics (specific, on the ground) within each strategy. These strategies and tactics, intended to guide both short- and long-term planning and management, were required to be feasible with respect to budget and level of effort, and to be acceptable within current policies.

5.5.1 Adaptation Strategies and Tactics

Of the many adaptation options identified for forest vegetation in the Northern Rockies (Halofsky et al. 2017), the major ones are summarized in Table 5.3. Many of the adaptation options are focused on protecting forests from and building

Table 5.3 Climate change adaptation options and restoration potential for tree species in the Northern Rockies

Species	Primary adaptive tactics	Restoration potential	Additional management recommendations
Ponderosa pine	Restore fire to historically fire-dominated stands; reduce fuel loadings to mitigate uncharacteristic fire severities; use HRV to guide restoration treatments.	Moderate to high. Reintroduce fire in fire-excluded stands as the first step; then identify where to plant in the future.	Reduce Douglas-fir in fire-excluded stands; remove competition with thinning and prescribed burns; monitor lower treeline in SW Montana and central Idaho.
Douglas-fir	Reduce competition and increase vigor; maintain low stem density; replace Douglas-fir with other species where root disease is a concern; emphasize ponderosa pine in low-elevation dry forests.	Moderate to high. Mitigate effects of fire exclusion era as the first step; reintroduce fire if possible (difficult in cool, dry environment).	Change species composition on sites where root disease and soil moisture deficits will increase; focus planting on higher elevation, mesic sites.
Western larch	Restore declining larch stands; prioritize treatments on north aspects and ash-cap soils; reduce competition; manage larch intensively on xeric sites; reduce stand density.	Moderate to high in Western Rockies. Moderate in Central Rockies.	Remove shade-tolerant species using group selection and thinning; prioritize planting options on north slopes; use genetic stock with best adaptive traits for drought and moisture stress.
Western white pine	Promote propagation of genotypes with resistance to white pine blister rust	Moderate in Western Rockies. Low to moderate in Central Rockies.	Increase planting of genotypes that have resistance to blister rust; thin dense stands to increase vigor of young pines
Grand fir	Ensure landscape heterogeneity; ensure age-class structure is near HRV.	High in Western and Central Rockies.	Invest in restoration only if the species is declining locally.
Western redcedar	Ensure landscape heterogeneity; maintain age-class diversity.	High in Western and Central Rockies.	Invest in restoration only if the species is declining locally.
Western hemlock	Ensure landscape heterogeneity; maintain age-class diversity.	High in Western and Central Rockies.	Invest in restoration only if the species is declining locally.
Lodgepole pine	Manage for mixed age classes and successional stages that approximate HRV.	Moderate to high.	Allow wildfires to burn where possible.

(continued)

Table 5.3 (continued)

Species	Primary adaptive tactics	Restoration potential	Additional management recommendations
Limber pine	Promote white pine blister rust resistance while preserving genetic diversity; monitor mortality rates and distribution; determine effects of fire exclusion.	Low to moderate. Most actions should increase rust resistance in native populations.	Implement rust-resistance programs; identify superior genotypes; collect cones and determine rust resistance; map limber pine populations to identify stands established before and after fire exclusion.
Subalpine fir	Use wildfire suppression to reduce species loss locally.	High.	Invest in restoration only if the species is declining locally.
Engelmann spruce	Use wildfire suppression to reduce species loss locally; plant selectively where populations are declining.	Mostly high, but moderate in low-elevation wet sites.	Invest in restoration only if the species is declining locally.
Mountain hemlock	Use wildfire suppression to reduce species loss locally.	Moderate to high in Western and Central Rockies.	Monitor to ensure the species is not locally extirpated.
Alpine larch	Preserve genetic diversity by collecting and storing seed.	Low to moderate.	Monitor changes in alpine larch populations.
Whitebark pine	Follow strategies in Keane et al. (2012); promote propagation of genotypes with resistance to white pine blister rust; conserve genetic diversity; prioritize treatments at high elevation.	Low to moderate because of stress imposed by blister rust.	Protect rust-resistant, high-vigor trees; implement prescribed fire and mechanical cuttings to reduce competition; plant and direct-seed rust-resistant seedlings on burns and treated areas; use hardy, drought-tolerant seedlings.
Quaking aspen	Restore quasi-historical fire regimes; prioritize areas where aspen already exists, even if at lower than historical levels.	Moderate.	Plant aspen where now absent but once existed; ensure diversity of age classes and seral stages across landscapes.
Cottonwood	Encourage high variability in streamflows to increase seedling establishment; reduce competition.	Moderate to high.	Prioritize the most mesic sites first; allow fire to burn in areas that are not too dense; remove competing conifers.
Green ash	Reduce grazing; use fire suppression and planting to promote ash populations in areas with low populations.	High in Eastern Rockies and Grassland.	Plant ash in recently burned areas where it recently existed.

resilience to severe disturbance, primarily wildfire. For example, promoting disturbance-resilient forest structure and species is a key adaptation strategy that guides management of vegetation and other resource areas in the Northern Rockies. Thinning and prescribed fire can be used to reduce forest density and promote disturbance-resilient species. Disturbance-resilient species can also be planted. Managers recognize the importance of promoting and planting site-adapted species, specifically western larch and western white pine on moist sites, ponderosa pine on dry sites, Douglas-fir on dry sites, and lodgepole pine on sites that are difficult to regenerate.

Preparing for disturbance will also be important in a changing climate. Tree regeneration after severe fire may be more limited in the future if drought frequency increases. Promoting legacy trees of disturbance-resilient species may help to increase postfire regeneration. Managers may also want to increase seed collection and ensure that adequate nursery stock is available for post-disturbance planting.

Promoting species diversity, genetic diversity, and landscape diversity is also a critical adaptation strategy. Increasing diversity is a "hedge your bets" strategy that reduces risk of major forest loss. Areas with low species and genetic diversity will probably be more susceptible to stressors associated with climate change, so promoting species and genetic diversity, through plantings and in thinning treatments, will increase forest resilience to changing climate. Promoting landscape heterogeneity, in terms of species and structure, will also increase resilience to wildfire, insects, and disease.

Managers identified several ways to increase knowledge and manage in the face of uncertainty. Implementation of an adaptive management framework can help address uncertainty and adjust management over time. In the context of climate change adaptation, adaptive management involves: definition of management goals, objectives and timeframes, analyzing vulnerabilities, determining priorities, developing adaptation strategies and tactics, implementing plans and projects, and monitoring, reviewing, and adjusting (Millar et al. 2014). Development of a consistent monitoring framework that can capture ecosystem changes with shifting climate is a key component of the adaptive management framework. For example, tracking tree species regeneration and distribution will help managers determine how species are responding to climatic changes and how to adjust management accordingly (e.g., guidelines for planting). Integration between research and management and across resource areas (e.g., forest management and wildlife) will also be needed in implementation of the adaptive management framework to ensure that management approaches do not conflict (e.g., which effects will a particular thinning treatment have on wildlife?).

Managers also identified adaptation strategies and tactics to maintain particular species or community types of concern. For example, climate change will probably lead to increased whitebark pine mortality through white pine blister rust, mountain pine beetle activity, and wildfire. To promote resilient whitebark pine communities, managers may want to focus restoration efforts on sites less likely to be affected by climate change (refugia). A variety of management strategies can be implemented to promote whitebark pine, including fire management, planting at lower elevations, and removing other dominant species (e.g., lodgepole pine, spruce, and fir). Genetically selected seedlings can also be planted to promote blister rust resistance.

Because stressors associated with climate change will be spatially pervasive, it will be important for agencies to coordinate and work across boundaries. Agencies can coordinate by aligning budgets and priorities for programs of work, communicating about projects adjacent to other lands, and working across boundaries to maintain roads, trails, and access that will be more frequently impacted by fire and flood events under changing climate.

Acknowledgments We thank Art Zack, Megan Strom, Chris Schnepf, Paul Zambino, and Kevin Greenleaf for technical reviews. We thank Gregg DiNitto, Paul Zambino, Blakey Lockman, Marcus Jackson, Joel Egan, Sandra Kegley, and Brytten Steed for contributions to the disturbance sections.

References

Alberto, F. J., Aitken, S. N., Alía, R., et al. (2013). Potential for evolutionary responses to climate change–evidence from tree populations. *Global Change Biology, 19*, 1645–1661.

Alexander, R. R., & Shepperd, W. D. (1990). *Picea engelmannii* Parry ex Engelm. Engelmann spruce. In R. M. Burns & B. H. Honkala (Eds.), *Silvics of North America: Volume 1 conifers* (pp. 403–444). Washington, DC: U.S. Forest Service.

Alexander, R. R., Shearer, R. C., & Shepperd, W. D. (1990). Abies lasiocarpa (Hook.) Nutt. Subalpine Fir. In R. M. Burns & B. H. Honkala (Eds.), *Silvics of North America: Volume 1 conifers* (pp. 149–166). Washington, DC: USDA Forest Service.

Allen, C. D., Macalady, A. K., Chenchouni, H., et al. (2010). A global overview of drought and heat-induced tree mortality reveals emerging climate change risks for forests. *Forest Ecology and Management, 259*, 660–684.

Arno, S. F. (1990). *Larix lyallii* Parl. alpine larch. In R. M. Burns & B. H. Honkala (Eds.), *Silvics of North America: Volume 1 conifers* (pp. 330–347). Washington, DC: U.S. Forest Service.

Arno, S. F. (2001). Community types and natural disturbance processes. In D. F. Tomback, S. F. Arno, R. E. Keane, & R.E. (Eds.), *Whitebark pine communities: Ecology and restoration* (pp. 74–89). Washington, DC: Island Press.

Arno, S. F., & Habeck, J. R. (1972). Ecology of alpine larch (*Larix lyallii* Parl.) in the Pacific Northwest. *Ecological Monographs, 42*, 417–450.

Aston, I. W. (2010). *Observed and projected ecological response to climate change in the Rocky Mountains and Upper Columbia Basin: A synthesis of current scientific literature* (Natural Resource Report NPS/ROMN/NPR—2010/220). Fort Collins: National Park Service.

Bartos, D. L. (1978). Modeling plant succession in aspen ecosystems. In D. N. Hyder (Ed.), *Proceedings of the First International Rangeland Congress* (pp. 208–211). Denver: Society for Range Management.

Baumgartner, D. M., Lotan, J. E., & Tonn, J. R. (1994). *Interior cedar-hemlock-white pine forests: Ecology and management*. Pullman: Washington State University Extension Service.

Bell, D. M., Bradford, J. B., & Lauenroth, W. K. (2014). Early indicators of change: Divergent climate envelopes between tree life stages imply range shifts in the western United States. *Global Ecology and Biogeography, 23*, 168–180.

Bentz, B. J., Régnière, J., Fettig, C. J., et al. (2010). Climate change and bark beetles of the western United States and Canada: Direct and indirect effects. *Bioscience, 60*, 602–613.

Beschta, R. L., & Ripple, W. J. (2005). Rapid assessment of riparian cottonwood recruitment: Middle Fork John Day River, Northeastern Oregon. *Ecological Restoration, 23*, 150–156.

Bigler, C., Kulakowski, D., & Veblen, T. T. (2005). Multiple disturbance interactions and drought influence fire severity in Rocky Mountain subalpine forests. *Ecology, 86*, 3018–3029.

Bollenbacher, B. (2012). *Characteristics of primary tree species in the Northern Region*. Missoula: U.S. Forest Service, Northern Region.

Brandt, J. P., Cerezke, H. F., Mallet, K. I., et al. (2003). Factors affecting trembling aspen (*Populus tremuloides* Michx.) health in Alberta, Saskatchewan, and Manitoba, Canada. *Forest Ecology and Management, 178*, 287–300.

Broadhurst, L. M., Lowe, A., Coates, D. J., et al. (2008). Seed supply for broadscale restoration: Maximizing evolutionary potential. *Evolutionary Applications, 1*, 587–597.

Brown, J. K. (1996). Fire effects on aspen and cottonwood. In: *Aspen and cottonwood in the Blue Mountains workshop*. La Grande: Blue Mountains Natural Resources Institute.

Brunelle, A., Whitlock, C., Bartlein, P., & Kipfmueller, K. (2005). Holocene fire and vegetation along environmental gradients in the Northern Rocky Mountains. *Quaternary Science Reviews, 24*, 2281–2300.

Burns, R. M., & Honkala, B. H. (1990). *Silvics of North America: Volume one, conifers*. Washington, DC: U.S. Forest Service.

Callaway, R., Sala, A., Keane, R. E., (1998). *Replacement of whitebark pine by subalpine fir: Consequences for stand carbon, water, and nitrogen cycles*. In USDA forest service, Fire Sciences Laboratory (p. 31). Missoula.

Carlson, C. E., Arno, S. F., & Menakis, J. (1990). Hybrid larch of the Carlton Ridge Research Natural Area in Western Montana. *Natural Areas Journal, 10*, 134–139.

Carroll, A. L., Taylor, S. W., Régnière, J., & Safranyik, L. (2003). Effects of climate change on range expansion by the mountain pine beetle in British Columbia. In *Mountain pine beetle symposium: Challenges and solutions* (pp. 223–231). Victoria: Natural Resources Canada/Canadian Forest Service.

Chhin, S., Hogg, E. T., Lieffers, V. J., & Huang, S. (2008). Potential effects of climate change on the growth of lodgepole pine across diameter size classes and ecological regions. *Forest Ecology and Management, 256*, 1692–1703.

Chmura, D. J., Anderson, P. D., Howe, G. T., et al. (2011). Forest responses to climate change in the Northwestern United States: Ecophysiological foundations for adaptive management. *Forest Ecology and Management, 261*, 1121–1142.

Clark, J. S., Carpenter, S. R., Barber, M., et al. (2001). Ecological forecasts: An emerging imperative. *Science, 293*, 657–660.

Cohn, J. S., Di Stefano, J., Christie, F., et al. (2015). How do heterogeneity in vegetation types and post-fire age-classes contribute to plant diversity at the landscape scale? *Forest Ecology and Management, 346*, 22–30.

Coops, N. C., & Waring, R. H. (2011). A process-based approach to estimate lodgepole pine (*Pinus contorta* Dougl.) distribution in the Pacific Northwest under climate change. *Climatic Change, 105*, 313–328.

Creeden, E. P., Hicke, J. A., & Buotte, P. C. (2014). Climate, weather, and recent mountain pine beetle outbreaks in the western United States. *Forest Ecology and Management, 312*, 239–251.

Cumming, S. G., & Burton, P. J. (1996). Phenology-mediated effects of climatic change on some simulated British Columbia forests. *Climatic Change, 34*, 213–222.

Cushman, S. A., McKenzie, D., Peterson, D. L., et al. (2007). *Research agenda for integrated landscape modeling* (General Technical Report RMRS-GTR-194). Fort Collins: U.S. Forest Service, Rocky Mountain Research Station.

Devine, W., Aubry, C., Bower, A., et al. (2012). *Climate change and forest trees in the Pacific Northwest: A vulnerability assessment and recommended actions for national forests*. Olympia: U.S. Forest Service/Pacific Northwest Region.

Dickman, A., & Cook, S. (1989). Fire and fungus in a mountain hemlock forest. *Canadian Journal of Botany, 67*, 2005–2016.

Falk, D. A., Miller, C., McKenzie, D., & Black, A. E. (2007). Cross-scale analysis of fire regimes. *Ecosystems, 10*, 809–823.

Ferguson, D. E., & Johnson, F. D. (1996). *Classification of grand fir mosaic habitats* (General Technical Report INT-GTR-337). Ogden: U.S. Forest Service, Intermountain Research Station.

Fins, L., & Steeb, L. W. (1986). Genetic variation in allozymes of western larch. *Canadian Journal of Forest Research, 16*, 1013–1018.

Fins, L., Byler, J., Ferguson, D., et al. (2002). Return of the giants: Restoring white pine ecosystems by breeding and aggressive planting of blister rust-resistant white pines. *Journal of Forestry, 100*, 20–26.

Flannigan, M. D., Amiro, B. D., Logan, K. A., et al. (2005). Forest fires and climate change in the 21st century. *Mitigation and Adaptation Strategies for Global Change, 11*, 847–859.

Flannigan, M. D., Krawchuk, M. A., de Groot, W. J., et al. (2009). Implications of changing climate for global wildland fire. *International Journal of Wildland Fire, 18*, 483–507.

Foiles, M. W., Graham, R. T., & Olson, D. F. (1990). *Abies grandis* (Dougl. ex D. Don) Lindl. grand fir. In R. M. Burns & B. H. Honkala (Eds.), *Silvics of North America: Volume 1 conifers* (pp. 132–155). Washington, DC: U.S. Forest Service.

Frey, B. R., Lieffers, V. J., Hogg, E. H., & Landhäusser, S. M. (2004). Predicting landscape patterns of aspen dieback: Mechanisms and knowledge gaps. *Canadian Journal of Forest Research, 34*, 1379–1390.

Fulé, P. Z., Cocke, A. E., Heinlein, T. A., & Covington, W. W. (2004). Effects of an intense prescribed forest fire: Is it ecological restoration? *Restoration Ecology, 12*, 220–230.

Funk, J., & Saunders, S. (2014). *Rocky mountain forests at risk: Confronting climate-driven impacts from insects, wildfires, heat, and drought*. Cambridge, MA: Union of Concerned Scientists.

Graham, R. T. (1990). *Pinus monticola* Dougl. ex D. Don western white pine. In R. M. Burns & B. H. Honkala (Eds.), *Silvics of North America: Volume 1 conifers* (pp. 348–353). Washington, DC: U.S. Forest Service.

Gray, L. K., & Hamann, A. (2013). Tracking suitable habitat for tree populations under climate change in western North America. *Climatic Change, 117*, 289–303.

Halofsky, J. E., Peterson, D. L., Dante, S. K., et al. (2017). *Climate change vulnerability and adaptation in the Northern Rocky Mountains*. (General Technical Report RMRS-GTR-xxx). Fort Collins: U.S. Forest Service, Rocky Mountain Research Station. In press.

Hamann, A., & Wang, T. (2006). Potential effects of climate change on ecosystem and tree species distribution in British Columbia. *Ecology, 87*, 2733–2786.

Hamrick, J. L. (2004). Response of forest trees to global environmental changes. *Forest Ecology and Management, 197*, 323–335.

Hansen, A. J., Neilson, R. P., Dale, V. H., et al. (2001). Global change in forests: Responses of species, communities, and biomes: Interactions between climate change and land use are projected to cause large shifts in biodiversity. *Bioscience, 51*, 765–779.

Ibañez, I., Clark, J. S., LaDeau, S., & Hille Ris Lambers, J. (2007). Exploiting temporal variability to understand tree recruitment response to climate change. *Ecological Monographs, 77*, 163–177.

Intergovernmental Panel on Climate Change (IPCC). (2007). *Climate change 2007—the physical science basis. Climate Change 2007 working group I contribution to the fourth assessment report of the IPCC*. New York: Cambridge University Press.

Ireland, K. B., Moore, M. M., Fulé, P. Z., et al. (2014). Slow lifelong growth predisposes *Populus tremuloides* trees to mortality. *Oecologia, 175*, 847–859.

Iverson, L. R., & McKenzie, D. (2013). Tree-species range shifts in a changing climate: Detecting, modeling, assisting. *Landscape Ecology, 28*, 879–889.

Iverson, L. R., & Prasad, A. M. (2002). Potential redistribution of tree species habitat under five climate change scenarios in the eastern US. *Forest Ecology and Management, 155*, 205–222.

Jackson, M., Gannon, A., Kearns, H., & Kendall, K. C. (2010). *Current status of limber pine in Montana* (Report 10-06). Missoula: U.S. Forest Service, Northern Region.

Johnstone, J. F., & Chapin, F. S. (2003). Non-equilibrium succession dynamics indicate continued northern migration of lodgepole pine. *Global Change Biology, 9*, 1401–1409.

Joyce, L. A., & Birdsey, R.A. (2000). *The impact of climate change on America's forests: A technical document supporting the 2000 USDA Forest Service RPA assessment* (General Technical Report RMRS-GTR-59). Fort Collins: U.S. Forest Service, Rocky Mountain Research Station.

Joyce, L. A., Blate, G. M., Littell, J. S., et al. (2008). National forests. In S. H. Julius & W. West (Eds.), *Preliminary review of adaptation options for climate-sensitive ecosystmes and resources. Ch. 4*. Washington, DC: U.S. Environmental Protection Agency.

Joyce, L. A., Running, S. W., Breshears, D. D., et al. (2014). Forests. In J. M. Melillo, T. C. Richmond, & G. W. Yohe (Eds.), *Climate change impacts in the United States: Third National Climate Assessment. Ch. 7* (pp. 175–194). Washington, DC: U.S. Global Change Research Program.

Jump, A. S., & Peñuelas, J. (2005). Running to stand still: Adaptation and the response of plants to rapid climate change. *Ecology Letters, 8*, 1010–1020.

Kappelle, M., Van Vuuren, M. M. I., & Baas, P. (1999). Effects of climate change on biodiversity: A review and identification of key research issues. *Biodiversity and Conservation, 8*, 1383–1397.

Keane, R. E. (2012). Creating historical range of variation (HRV) time series using landscape modeling: Overview and issues. In J. A. Wiens, G. D. Hayward, H. S. Safford, C. Giffen, & C. (Eds.), *Historical environmental variation in conservation and natural resource management* (pp. 113–128). Hoboken: John Wiley and Sons.

Keane, R. E. (2013). Disturbance regimes and the historical range of variation in terrestrial ecosystems. In A. L. Simon (Ed.), *Encyclopedia of biodiversity* (2nd ed., pp. 568–581). Waltham: Academic Press.

Keane, R. E., & Parsons, R. (2010). Restoring whitebark pine forests of the northern Rocky Mountains, USA. *Ecological Restoration, 28*, 56–70.

Keane, R. E., Morgan, P., & Menakis, J. P. (1994). Landscape assessment of the decline of whitebark pine (*Pinus albicaulis*) in the Bob Marshall Wilderness Complex, Montana, USA. *Northwest Science, 68*, 213–229.

Keane, R. E., Ryan, K. C., & Running, S. W. (1996). Simulating effects of fire on northern Rocky Mountain landscapes with the ecological process model Fire-BGC. *Tree Physiology, 16*, 319–331.

Keane, R. E., Veblen, T., Ryan, K. C., et al. (2002). The cascading effects of fire exclusion in the Rocky Mountains. In J. Baron (Ed.), *Rocky Mountain futures: An ecological perspective* (pp. 133–153). Washington, DC: Island Press.

Keane, R. E., Cary, G., Davies, I. D., et al. (2004). A classification of landscape fire succession models: Spatially explicit models of fire and vegetation dynamic. *Ecological Modelling, 256*, 3–27.

Keane, R. E., Agee, J., Fulé, P., et al. (2008). Ecological effects of large fires in the United States: Benefit or catastrophe. *International Journal of Wildland Fire, 17*, 696–712.

Keane, R. E., Hessburg, P. F., Landres, P. B., & Swanson, F. J. (2009). A review of the use of historical range and variation (HRV) in landscape management. *Forest Ecology and Management, 258*, 1025–1037.

Keane, R. E., Tomback, D. F., Aubry, C. A., et al. (2012). *A range-wide restoration strategy for whitebark pine forests* (General Technical Report RMRS-GTR-279). Fort Collins: U.S. Forest Service, Rocky Mountain Research Station.

Keane, R. E., Loehman, R., Clark, J., et al. (2015a). Exploring interactions among multiple disturbance agents in forest landscapes: Simulating effects of fire, beetles, and disease under climate change. In A. H. Perera, B. R. Surtevant, & L. J. Buse (Eds.), *Simulation modeling of forest landscape disturbances* (pp. 201–231). New York, NY: Springer.

Keane, R. E., McKenzie, D., Falk, D. A., et al. (2015b). Representing climate, disturbance, and vegetation interactions in landscape models. *Ecological Modelling, 309–310*, 33–47.

Kirilenko, A. P., & Sedjo, R. A. (2007). Climate change impacts on forestry. *Proceedings of the National Academy of Sciences, USA, 104*, 19697–19702.

Landres, P. B., Morgan, P., & Swanson, F. J. (1999). Overview and use of natural variability concepts in managing ecological systems. *Ecological Applications, 9*, 1179–1188.

Lanner, R. M. (1980). Avian seed dispersal as a factor in the ecology and evolution of limber and whitebark pines. In *Sixth North American Forest Biology Workshop* (pp. 15–47). Alberta: University of Alberta.

Lawler, J. J., White, D., Neilson, R. P., & Blaustein, A. R. (2006). Predicting climate-induced range shifts: Model differences and model reliability. *Global Change Biology, 12*, 1568–1584.

Ledig, F. T., & Kitzmiller, J. H. (1992). Genetic strategies for reforestation in the face of global climate change. *Forest Ecology and Management., 50*, 153–169.

Lesica, P. (2001). Recruitment of *Fraxinus pennsylvanica* (Oleaceae) in eastern Montana woodlands. *Madrono, 48*, 286–292.

Lesica, P. (2009). Can regeneration of green ash (*Fraxinus pensylvanica*) be restored in declining woodlands in eastern Montana? *Rangeland Ecology and Management, 62*, 564–571.

Liang, Y., He, H. S., Wang, W. J., et al. (2015). The site-scale processes affect species distribution predictions of forest landscape models. *Ecological Modelling, 300*, 89–101.

Loehman, R. A., Clark, J. A., & Keane, R. E. (2011). Modeling effects of climate change and fire management on western white pine (*Pinus monticola*) in the Northern Rocky Mountains, USA. *Forests, 2*, 832–860.

Loehman, R. A., Reinhardt, E., & Riley, K. L. (2014). Wildland fire emissions, carbon, and climate: Seeing the forest and the trees—a cross-scale assessment of wildfire and carbon dynamics in fire-prone, forested ecosystems. *Forest Ecology and Management, 317*, 9–19.

Logan, J. A., & Powell, J. A. (2001). Ghost forests, global warming, and the mountain pine beetle (*Coleoptera: Scolytidae*). *American Entomologist, 47*, 160–173.

Logan, J. A., Regniere, J., & Powell, J. A. (2003). Assessing the impacts of global warming on forest pest dynamics. *Frontiers in Ecology and the Environment, 1*, 130–137.

Lotan, J. E., & Critchfield, W. B. (1990). *Pinus contorta* Dougl. ex. Loud. lodgepole pine. In R. M. Burns & B. H. Honkala (Eds.), *Silvics of North America: Volume 1 conifers* (pp. 648–666). Washington, DC: U.S. Forest Service.

Marchetti, S. B., Worrall, J. J., & Eager, T. (2011). Secondary insects and diseases contribute to sudden aspen decline in southwestern Colorado, USA. *Canadian Journal of Forestry Research, 41*, 2315–2325.

Marlon, J. R., Bartlein, P. J., Walsh, M. K., et al. (2009). Wildfire responses to abrupt climate change in North America. *Proceedings of the National Academy of Sciences, USA, 106*, 2519–2524.

McKenzie, D., Peterson, D. L., & Littell, J. S. (2009). Global warming and stress complexes in forests of western North America. In A. Bytnerowicz, M. J. Arbaugh, A. R. Riebau, & C. Andersen (Eds.), *Wildland fires and air pollution* (pp. 317–337). The Hague: Elsevier.

McKenzie, D., Miller, C., & Falk, D. A. (Eds.). (2011). *The landscape ecology of fire*. Dordrecht: Springer.

McKenzie, D., Shankar, U., Keane, R. E., et al. (2014). Smoke consequences of new wildfire regimes driven by climate change. *Earth's Future, 2*, 35–39.

Means, J. E. (1990). *Tsuga mertensiana* (Bong.) Carr. mountain hemlock. In R. M. Burns & B. H. Honkala (Eds.), *Silvics of North America: Volume 1 conifers* (pp. 1318–1332). Washington, DC: U.S. Forest Service.

Millar, C. I., Swanston, C. W., & Peterson, D. L. (2014). Adapting to climate change. In D. L. Peterson, J. M. Vose, & T. Patel-Weynand (Eds.), *Climate change and United States forests* (pp. 183–222). Dordrecht: Springer.

Miller, J. D., Safford, H. D., Crimmins, M., & Thode, A. E. (2009). Quantitative evidence for increasing forest fire severity in the Sierra Nevada and Southern Cascade Mountains, California and Nevada, USA. *Ecosystems, 12*, 16–32.

Milner, K. S. (1992). Site index and height growth curves for ponderosa pine, western larch, lodgepole pine, and Douglas-fir in western Montana. *Western Journal of Applied Forestry, 7*, 9–14.

Minore, D. (1979). *Comparative autecological characteristics of northwestern tree species: A literature review* (General Technical Report PNW-GTR-087). Portland, OR.: U.S. Forest Service, Pacific Northwest Forest and Range Experiment Station.

Minore, D. (1990). *Thuja plicata* Donn ex D. Don western red cedar. In R. M. Burns & B. H. Honkala (Eds.), *Silvics of North America: Volume 1 conifers* (pp. 1249–1267). Washington, DC: U.S. Forest Service.

Mitton, J. B., & Grant, M. C. (1996). Genetic variation and the natural history of quaking aspen. *Bioscience, 46*, 25–31.

Morales, J. M., Mermoz, M., Gowda, J. H., & Kitzberger, T. (2015). A stochastic fire spread model for north Patagonia based on fire occurrence maps. *Ecological Modelling, 300*, 73–80.

Nigh, G. (2014). Mitigating the effects of climate change on lodgepole pine site height in British Columbia, Canada, with a transfer function. *Forestry, 87*, 377–387.

Nitschke, C. R., & Innes, J. L. (2008). A tree and climate assessment tool for modelling ecosystem response to climate change. *Ecological Modelling, 210*, 263–277.

Peterson, D. W., & Peterson, D. L. (2001). Mountain hemlock growth responds to climatic variability at annual and decadal scales. *Ecology, 82*, 3330–3345.

Peterson, D. L., Vose, J. M., & Patel-Weynand, T. (2014). *Climate change and United States forests*. Dordrecht: Springer.

Pfister, R. D., Kovalchik, B. L., Arno, S. F., & Presby, R. C. (1977). *Forest habitat types of Montana* (General Technical Report INT-GTR-34). Ogden: U.S. Forest Service, Intermountain Forest and Range Experiment Station.

Rehfeldt, G. E. (1994). Evolutionary genetics, the biological species, and the ecology of the cedar-hemlock forests. In D. M. Baumgartner, J. E. Lotan, & J. R. Tonn (Eds.), *Interior cedar-hemlock-white pine forests: Ecology and management* (pp. 91–100). Spokane: Washington State University, Cooperative Extension.

Rehfeldt, G. E. (1995). Genetic variation, climate models and the ecological genetics of *Larix occidentalis. Forest Ecology and Management, 78*, 21–37.

Rehfeldt, G. E., & Jaquish, B. C. (2010). Ecological impacts and management strategies for western larch in the face of climate-change. *Mitigation and Adaptation Strategies for Global Change, 15*, 283–306.

Restaino, C. M., Peterson, D. L., & Littell, J. S. (2016). Increased water deficit decreases Douglas-fir growth throughout western US forests. *Proceedings of the National Academy of Sciences, USA, 113*, 9557–9562.

Retzlaff, M. L., Leirfallom, S. B., & Keane, R. E. (2016). *A 20-year reassessment of the health and status of whitebark pine forests in the Bob Marshall Wilderness Complex, Montana* (Research Note RMRS-RN-73). Fort Collins: U.S. Forest Service, Rocky Mountain Research Station.

Riggs, R. A., Keane, R. E., Cimon, N., et al. (2015). Biomass and fire dynamics in a temperate forest-grassland mosaic: Integrating multi-species herbivory, climate, and fire with the FireBGCv2/GrazeBGC system. *Ecologial Modelling, 296*, 57–78.

Romme, W. H., Turner, M. G., Gardner, R. H., et al. (1997). A rare episode of sexual reproduction in aspen (*Populus tremuloides* Michx.) following the 1988 Yellowstone fires. *Natural Areas Journal, 17*, 17–25.

Rood, S. B., Braatne, J. H., & Hughes, F. M. (2003). Ecophysiology of riparian cottonwoods: Stream flow dependency, water relations and restoration. *Tree Physiology, 23*, 1113–1124.

Ryan, K. C., & Reinhardt, E. D. (1988). Predicting postfire mortality of seven western conifers. *Canadian Journal of Forest Research, 18*, 1291–1297.

Ryan, M. G., Gower, S. T., Hubbard, R. M., et al. (1995). Woody tissue maintenance respiration of four conifers in contrasting climates. *Oecologia, 101*, 133–140.

Safranyik, L., Carroll, A., Régnière, J., et al. (2010). Potential for range expansion of mountain pine beetle into the boreal forest of North America. *The Canadian Entomologist, 142*, 415–442.

Sala, A., Peters, G. D., McIntyre, L. R., & Harrington, M. G. (2005). Physiological responses of ponderosa pine in western Montana to thinning, prescribed fire and burning season. *Tree Physiology, 25*, 339–348.

Schauer, A. J., Wade, B. K., & Sowell, J. B. (1998). Persistence of subalpine forest-meadow ecotones in the Gunnison Basin, Colorado. *Great Basin Naturalist, 58*, 273–281.

Schoettle, A. W., & Sniezko, R. A. (2007). Proactive intervention to sustain high-elevation pine ecosystems threatened by white pine blister rust. *Journal of Forest Research, 12*, 327–336.

Schumacher, S., Reineking, B., Sibold, J., & Bugmann, H. (2006). Modeling the impact of climate and vegetation on fire regimes in mountain landscapes. *Landscape Ecology, 21*, 539–554.

Shafer, S. L., Bartlein, P. J., & Thompson, R. S. (2001). Potential changes in the distributions of western north America tree and shrub taxa under future climate scenarios. *Ecosystems, 4*, 200–215.

Shearer, R. C., & Schimidt, W. C. (1970). *Natural regeneration in ponderosa pine forests of western Montana* (Research Paper INT-RP-86). Ogden: U.S. Forest Service, Intermountain Forest and Range Experiment Station.

Smithwick, E. A. H., Ryan, M. G., Kashian, D. M., et al. (2009). Modeling the effects of fire and climate change on carbon and nitrogen storage in lodgepole pine (*Pinus contorta*) stands. *Global Change Biology, 15*, 535–548.

Spittlehouse, D. L., & Stewart, R. B. (2003). Adaptation to climate change in forest management. *British Columbia Journal of Ecosystems and Management, 4*, 1–11.

St. Clair, J. B., & Howe, G. T. (2007). Genetic maladaptation of coastal Douglas-fir seedlings to future climates. *Global Change Biology, 13*, 1441–1454.

Steele, R. (1990). *Pinus flexilis* James limber pine. In R. M. Burns & B. H. Honkala (Eds.), *Silvics of North America: Volume 1 conifers* (pp. 348–353). Washington, DC: U.S. Forest Service.

Stout, D. L., & Sala, A. (2003). Xylem vulnerability to cavitation in *Pseudotsuga menziesii* and *Pinus ponderosa* from contrasting habitats. *Tree Physiology, 23*, 43–50.

Swanston, C., & Janowiak, M. (Eds.). (2012). *Forest adaptation resources: Climate change tools and approaches for land managers* (General Technical Report NRS-GTR-87). Newtown Square: U.S. Forest Service, Northern Research Station.

Taylor, J., & Sturdevant, N. (1998). *Limber pine mortality on the Lewis and Clark National Forest, Montana* (Forest Health Protection Report 98-5). Missoula: U.S. Forest Service, Nothern Region.

Temperli, C., Bugmann, H., & Elkin, C. (2013). Cross-scale interactions among bark beetles, climate change, and wind disturbances: A landscape modeling approach. *Ecological Monographs, 83*, 383–402.

Tomback, D., Arno, S. F., & Keane, R. E. (2001). *Whitebark pine communities: Ecology and restoration*. Washington, DC: Island Press.

Turner, M. G., Romme, W. H., Gardner, R. H., et al. (1993). A revised concept of landscape equilibrium: Disturbance and stability on scaled landscapes. *Landscape Ecology, 8*, 213–227.

Urban, D. L., Harmon, M. E., & Halpern, C. B. (1993). Potential response of Pacific Northwestern forests to climatic change, effects of stand age and initial composition. *Climatic Change, 23*, 247–266.

Wang, T., Hamann, A., Yanchuk, A., et al. (2006). Use of response functions in selecting lodgepole pine populations for future climates. *Global Change Biology, 12*, 2404–2416.

Whited, D. C., Lorang, M. S., Harner, M. J., et al. (2007). Climate, hydrologic disturbance, and succession: Drivers of floodplain pattern. *Ecology, 88*, 940–953.

Whitlock, C. (1993). Postglacial vegetation and climate of Grand Teton and southern Yellowstone National Parks. *Ecological Monographs, 63*, 173–198.

Whitlock, C. (2004). Forests, fire and climate. *Nature, 432*, 28–29.

Whitlock, C., & Bartlein, P. J. (1993). Spatial variations of holocene climatic change in the Yellowstone region. *Quaternary Research, 39*, 231–238.

Whitlock, C., Shafer, S. L., & Marlon, J. (2003). The role of climate and vegetation change in shaping past and future fire regimes in the northwestern US and the implications for ecosystem management. *Forest Ecology and Management, 178*, 5–21.

Woods, A. J., Heppner, D., Kope, H. H., et al. (2010). Forest health and climate change: A British Columbia perspective. *The Forestry Chronicle, 86*, 412–422.

Woodward, A., Schreiner, E. G., & Silsbee, D. G. (1995). Climate, geography, and tree establishment in subalpine meadows of the Olympic Mountains, Washington, USA. *Arctic and Alpine Research, 27*, 217–225.

Chapter 6
Effects of Climate Change on Rangeland Vegetation in the Northern Rockies

Matt C. Reeves, Mary E. Manning, Jeff P. DiBenedetto, Kyle A. Palmquist, William K. Lauenroth, John B. Bradford, and Daniel R. Schlaepfer

Abstract A longer growing season with climate change is expected to increase net primary productivity of many rangeland types, especially those dominated by grasses, although responses will depend on local climate and soil conditions. Elevated atmospheric carbon dioxide may increase water use efficiency and productivity of some species. In many cases, increasing wildfire frequency and extent will be damaging for big sagebrush and other shrub species that are readily killed by fire. The widespread occurrence of cheatgrass and other nonnatives facilitates frequent fire through annual fuel accumulation. Shrub species that sprout following fire may be quite resilient to increased disturbance, but may be outcompeted by more drought tolerant species over time.

Adaptation strategies for rangeland vegetation focus on increasing resilience of rangeland ecosystems, primarily through non-native species control and prevention. Ecologically based non-native plant management focuses on strategies to repair

---

M.C. Reeves (✉)
U.S. Forest Service, Rocky Mountain Research Station, Missoula, MT, USA
e-mail: mreeves@fs.fed.us

M.E. Manning
U.S. Forest Service, Northern Region, Missoula, MT, USA
e-mail: mmanning@fs.fed.us

J.P. DiBenedetto
U.S. Forest Service, Custer National Forest, Retired, Billings, MT, USA
e-mail: jp_dibenedetto@msn.com

K.A. Palmquist • W.K. Lauenroth
Department of Botany, University of Wyoming, Laramie, WY, USA
e-mail: kpalmqu1@uwyo.edu; wlauenro@uwyo.edu

J.B. Bradford
U.S. Geological Survey, Southwest Biological Science Center, Flagstaff, AZ, USA
e-mail: jbradford@usgs.gov

D.R. Schlaepfer
Department of Environmental Sciences, University of Basel, Basel, Switzerland
e-mail: daniel.schlaepfer@unibas.ch

damaged ecological processes that facilitate invasion, and seeding of desired natives can be done where seed availability and dispersal of natives are low. Proactive management to prevent establishment of non-native species is also critical (early detection-rapid response), including tactics such as weed-free policies, education of employees and the public, and collaboration among multiple agencies to control weeds. Livestock grazing can also be managed through the development of site-specific indicators that inform livestock movement guides and allow for maintenance and enhancement of plant health.

Keywords Rangelands • Vulnerability • Climate change • Nonnative plants • Adaptation • Sagebrush • Woodlands • Grasslands • Shrublands

6.1 Introduction

Rangelands, including grassland, shrubland, desert, alpine, and some woodland ecosystems, are dominated by grass, forb, or shrub species (Lund 2007). Rangelands occupy more than 26 million hectares in the Northern Rockies (Reeves and Mitchell 2011), producing forage for domestic and wild ungulates, providing critical habitat for numerous species such as greater sage-grouse (*Centrocercus urophasianus*), and providing many recreational opportunities.

Climate change, combined with residential development, energy development, and invasive (nonnative) species (e.g., cheatgrass, wild horses and burros), create a significant challenge for resource managers charged with ensuring sustainability of ecosystem services. The effects of climate change on rangelands have been studied less than effects on forests, but the effects of (past and future) human land-use activities on rangelands will probably exceed those of climate change, at least in the short term. This assessment focuses on regeneration success, response to disturbance (especially wildfire), and life history traits in rangelands, rather than on explicit estimates of future land-use change. The focus on life history traits combined with the concepts of resilience and resistance can help with understanding the effects of climate change. Resilience is the capacity of ecosystems to regain structure, processes, and function in response to disturbance (Holling 1973; Allen et al. 2005), whereas resistance is the capacity to retain these attributes in response to disturbance (Folke et al. 2004). These concepts are especially helpful for understanding establishment of nonnative plants and interactions between climate change stressors (Chambers et al. 2014), as demonstrated in Fig. 6.1, which shows that management for ecosystem services derived from rangelands will be most effective in mesic rangelands.

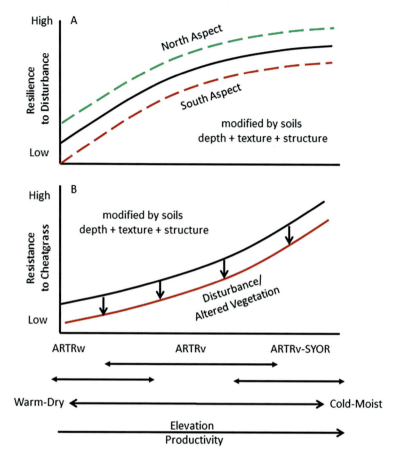

Fig. 6.1 Resilience to disturbance (**a**) and resistance to cheatgrass (**b**) over a typical temperature/precipitation gradient in cold desert (modified from Chambers et al. 2014). Dominant ecological sites range from Wyoming big sagebrush on warm, dry sites (*left*); to mountain big sagebrush on cool, moist sites (*middle*); to mountain big sagebrush and root-sprouting shrubs on cold, moist sites (*right*). Resilience increases along the temperature/precipitation gradient, influenced by site characteristics (e.g., aspect). Resistance also increases along the gradient, influenced by disturbances and management treatments that alter vegetation structure and composition. *ARTRw* Wyoming big sagebrush (*Artemisia tridentata* ssp. *wyomingensis*); *ARTRv* mountain big sagebrush (*A. tridentata* ssp. *vaseyana*), *SYOR* mountain snowberry (*Symphoricarpos oreophilus*)

6.2 Rangeland Vegetation

This assessment focuses on vegetation types and individual species for which sufficient information and data exist to make inferences about the effects of climate change. First, we reviewed the extent of rangelands in the Northern Rockies and generally confined analysis to U.S. Forest Service (USFS) rangelands (Reeves and Mitchell 2011) in the Northern Rockies. We determined that the complexity of rangeland vegetation, combined with a paucity of climate change effects studies, suggests that a grouping of individual vegetation types into classes is appropriate. Therefore, we assessed the following vegetation classes:

- Northern Great Plains—This vegetation class is broadly distributed, including a mixture of cool-season (C3) and warm-season (C4) grass species.
- Montane shrubs—Includes a broad diversity of shrub species, many of which are important for browsing by native ungulates.
- Montane grasslands—This relatively scarce vegetation class is dominated by cool-season species, often intermixed with forest vegetation.
- Sagebrush systems—Dominated by species in the genus *Artemisia*, this is a ubiquitous and iconic vegetation class in much of the western United States, providing critical wildlife habitat for many species, including greater sage-grouse.

Sagebrush systems dominated by big sagebrushes (Wyoming big sagebrush [*Artemisia tridentata* ssp. *wyomingensis*], mountain big sagebrush [*A. t.* ssp. *vaseyana*], and basin big sagebrush [*A. t.* ssp. *tridentata*]) have been widely studied, at least partially as a result of recent research on sage-grouse habitat. Therefore, inferences about the vulnerability of these sagebrush species to climate change is supported by relatively more information than for other species. Four sagebrush types were delineated for this assessment: big sagebrushes (Wyoming big sagebrush, basin big sagebush), low sagebrushes (low sagebrush [*A. arbuscula*], black sagebrush [*A. nova*]), sprouting sagebrushes (silver sagebrush [*A. cana*], three-tip sagebrush [*A. tripartita*]), and mountain big sagebrush.

Wyoming and basin big sagebrush types were aggregated because they have similar life histories, stature, and areal coverage in the Northern Rockies, and because they represent critical habitats for many species of animals. Basin big sagebrush occupies sites with relatively deeper soils that retain sufficient moisture for perennial bunchgrasses, suggesting these sites may be more resilient and resistant to a drier climate (Chambers et al. 2007). Silver sagebrush and three-tip sagebrush can resprout after fire, making them unique among the sagebrush species. Communities dominated by Wyoming big sagebrush are by far the most common and occupy the most area (Table 6.1), whereas the low sagebrush type occupies the least. Although basin and Wyoming sagebrush are common throughout the Northern Rockies, mountain big sagebrush communities occupy the greatest extent on lands managed by the USFS.

6.3 Management Issues

Fire regimes, improper grazing, and nonnative species are concerns for rangeland management in the Northern Rockies. Uncharacteristic fire regimes threaten most rangeland habitats, especially sagebrush steppe, across much of the western United States. On one hand, "too much fire" may affect the landscape relative to historical fire regimes, because many sagebrush habitats now have shortened fire return

Table 6.1 Area of rangeland vegetation classes in each Northern Rockies subregion

Subregion	Rangeland vegetation classes	Area *Hectares*	Proportion *Percent*
Western Rockies	Montane grasslands	241,531	34.4
	Montane shrubs	120,658	35.7
	Sagebrush systems	144,912	29.9
	Total	507,101	
Central Rockies	Montane grasslands	342,177	43.6
	Montane shrubs	70,407	18.6
	Sagebrush systems	205,334	37.8
	Total	617,918	
Eastern Rockies	Montane grasslands	297,751	13.5
	Montane shrubs	132,861	12.5
	Northern Great Plains (C3/C4 mix)	89,514	5.9
	Sagebrush systems	1,040,907	68.2
	Total	1,561,033	
Greater Yellowstone area	Montane grasslands	222,282	6.1
	Montane shrubs	150,336	8.5
	Northern Great Plains (C3/C4 mix)	18,554	0.7
	Sagebrush systems	2,140,008	84.7
	Total	2,531,180	
Grassland	Montane grasslands	543,840	1.8
	Montane shrubs	107,740	0.7
	Northern Great Plains (C3/C4 mix)	16,674,787	80.6
	Sagebrush systems	3,474,994	16.8
	Total	20,801,361	
All subregions total		26,018,593	

intervals, resulting in increasing dominance of nonnative invasive annual grasses that create fuel conditions that facilitate more frequent combustion (Chambers et al. 2007). On the other hand, fire exclusion has led to longer fire return intervals that may be responsible for conifer encroachment in montane grasslands (Arno and Gruell 1986) and higher elevation sagebrush habitats, especially those dominated by mountain big sagebrush (Heyerdahl et al. 2006) (Fig. 6.2).

The nonnative invasive species of greatest concern is cheatgrass (*Bromus tectorum*), although Japanese brome (*B. japonicus*) and leafy spurge (*Euphorbia esula*) are also problems in the Northern Great Plains. Distribution of cheatgrass has expanded greatly in the western half of the Northern Rockies (Ramakrishnan et al. 2006; Merrill et al. 2012), and it is likely that further expansion may be enhanced by elevated atmospheric CO_2 concentrations, increased soil disturbance, and increasing spring and winter temperatures (Chambers et al. 2014; Boyte et al. 2016; Bradley et al. 2016).

Improper grazing—the mismanagement of grazing that produces detrimental effects on vegetation or soil resources—can create additional stress in some range-

Fig. 6.2 Establishment of ponderosa pine (*Pinus ponderosa*) and other conifers in montane grassland dominated by rough fescue

lands, accelerating the annual grass invasion/fire cycle, especially in some sagebrush types, the Northern Great Plains, and montane grasslands. Fortunately, most U.S. rangelands are not improperly grazed to the point of degradation (Reeves and Mitchell 2011; Reeves and Baggett 2014), a generalization that is true for most rangelands in the Northern Rockies.

6.4 Assessing the Effects of Climate Change on Rangelands

Despite the lack of focused studies on the effects of climate change on rangeland vegetation and the large uncertainty of projected climates, there are a few elements of climate change that are increasingly recognized as potential outcomes. First, projected temperature increases (Chap. 2) are expected to increase evaporative demand and pose greater overall stress (Polley et al. 2013). Projected changes in precipitation patterns and increasing potential evapotranspiration could facilitate more frequent wildfires through the combined effects of early-season plant growth and the desiccating effects of warmer summers (Morgan et al. 2008). These changes will lead to drier soils, particularly in summer when plants are physiologically active (Polley et al. 2013; Bradford et al. 2014; Palmquist et al. 2016a, b). However, winter precipitation is projected to increase 10–20% in the Northern Rockies (Chap. 2), which may compensate for increasing droughts. In addition, higher atmospheric CO_2 may offset evaporative demand by increasing water use efficiency in plants. Relative to much of the rest of the United States, the Northern Rockies could experience an increase in annual net primary productivity (NPP) (Fig. 6.3), partially as a result of the likely increase in water use efficiency and increased growing season

6 Effects of Climate Change on Rangeland Vegetation in the Northern Rockies 103

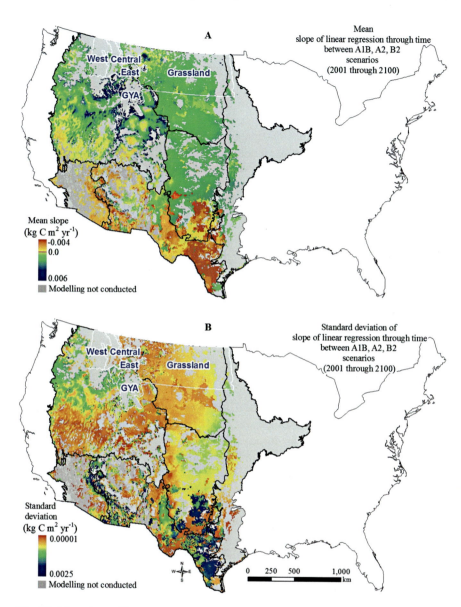

Fig. 6.3 Mean slope of linear regression for the net primary productivity trend for low (B2), moderate (A1B), and high (A2) emission scenarios (global climate models averaged: GCGM2, HadCM3, CSIRO, MK2, MIROC3.2) (**a**), and standard deviation of the mean slope of linear regression of the net primary productivity trend for the same scenarios (**b**) (From Reeves et al. 2014)

length (Reeves et al. 2014). Removal of growth limitations could result in significant changes in vegetation at higher elevations, such as the Greater Yellowstone Area. Higher NPP may seem counterintuitive because increased temperatures are associated with higher moisture stress and less favorable growing conditions. However, it is reasonable that high-elevation vegetation may experience increased production with increasing temperatures (Reeves et al. 2014), especially relatively mesic areas supporting mountain sagebrush. Increased atmospheric CO_2 concentrations may modify physiological growth processes in rangeland vegetation by enhancing water use efficiency, but response may not be consistent across all vegetation (Morgan et al. 2004b, 2011; Woodward and Kelly 2008).

Warmer winters and decreasing snowpack may have a significant effect on the distribution and abundance of different plant species. Minimum temperatures are expected to increase more than maximum temperatures, providing longer frost-free periods (Chap. 2). Warmer, wetter winters would favor early-season plant species and tap-rooted species that are able to access early-season soil water (Polley et al. 2013).

6.4.1 Montane Grasslands

Montane grasslands are associated with mountainous portions of the Northern Rockies, including the Palouse prairie and canyon grasslands of northern and central Idaho. Montane grasslands occur in intermountain valleys, foothills, and mountain slopes from low to relatively high elevation. They are dominated by cool-season (C3) grasses, many forbs, upland sedges, and scattered trees in some areas. Dominant species include bluebunch wheatgrass (*Pseudoroegneria spicata*), rough fescue (*Festuca campestris*), Idaho fescue (*F. idahoensis*), Sandberg bluegrass (*Poa secunda*), needle-and-thread (*Hesperostipa comata*), western wheatgrass (*Pascopyrum smithii*), prairie junegrass (*Koeleria macrantha*), western needlegrass (*Achnatherum nelsonii*), and Richardson's needlegrass *(A. richardsonii)*.

Most grasslands, particularly at lower elevations, are disturbed, fragmented, and often occupied by nonnative plant species. Improper livestock grazing, native ungulate herbivory, and nonnative plants are stressors in these grasslands. Lack of fire is also a stressor, because it can allow conifers to become established within grasslands (Arno and Gruell 1986; Heyerdahl et al. 2006). As conifer density increases with fire exclusion, grass cover declines, because most grassland species are shade intolerant. However, if fires become hotter and more frequent, there is an increased risk of mortality of native species and invasion by nonnative species (Ortega et al. 2012). As noted above, cheatgrass creates continuous fine fuels that are combustible by early summer; if fire occurs at this time, it can burn native perennial grasses before they have matured and set seed (Chambers et al. 2007; Bradley 2008). Several other nonnative species can also increase after fire, reducing native plant cover.

Nonnative plant species will probably expand in lower elevation grasslands as temperature increases, resistance to invasion decreases (Chambers et al. 2014), and disturbance increases (Bradley 2008). Drier conditions plus ungulate effects (grazing, browsing, hoof damage) may increase bare ground and possibly surface soil erosion. Low-elevation grasslands may have increasing dominance of more drought-tolerant species, such that cool-season species decline and warm-season species expand (Bachelet et al. 2001). However, elevated CO_2 favors C3 grasses and enhances biomass production, whereas warming favors C4 grasses (Morgan et al.

2004a, 2007). Therefore, a warmer and drier climate may allow C4 grasses to expand westward, displacing some C3 species. In general, a warmer climate with more frequent fires will favor increasing dominance of grasslands across the landscape, in some cases displacing shrublands and conifers that are burned too frequently to regenerate successively.

6.4.2 Montane Shrubs

Montane shrubs are associated with montane and subalpine forests, occurring as large patches within forested landscapes. Rocky Mountain maple *(Acer glabrum)*, oceanspray *(Holidiscus discolor)*, tobacco brush *(Ceanothus velutintis* var. *velutinus)*, Sitka alder *(Alnus viridus* subsp. *sinuata)*, thimbleberry *(Rubus parviflorus)*, chokecherry *(Prunus virginiana)*, serviceberry *(Amelanchier alnifolia)*, currant *(Ribes* spp.), snowberry *(Symphoricarpos albus)*, Scouler willow *(Salix scouleriana)*, and mountain ash *(Sorbus scopulina)* are common.

Montane shrubs persist in locations where disturbance kills aboveground stems, with subsequent sprouting from the root crown, rhizomes, and roots where adequate light and soil moisture are available. Fire exclusion, conifer establishment, and browsing by native and domestic animals are significant stressors. Most mesic shrubs are well adapted to frequent fire and can often compete well with associated conifers. However, even sprouting shrubs can sometimes be killed if fires are very hot and postfire weather is dry.

Drier soils and increased fire frequency may facilitate increasing dominance of more drought tolerant species such as rubber rabbitbrush *(Ericameria nauseosa)*, green rabbitbrush *(Chrysothamnus viscidiflorus)*, and spineless horsebrush *(Tetradymia canescens)*. Nonnative plant species may also expand into these communities, particularly following fire (Bradley 2008). Some mesic shrub species (e.g., Sitka alder, Rocky Mountain maple) may persist at higher elevations or in cooler, moister locations (e.g., north aspects, concavities).

6.4.3 Short Sagebrushes

Low sagebrush ecosystems cover only about 1% of total sagebrush habitat in the Northern Rockies, half of which is in the Western Rockies subregion. Low sagebrush sites have relatively low productivity, and are located between 1800 and 2700 m elevation in Montana and Idaho, occupying shallow claypan soils that restrict drainage and root growth. Black sagebrush is found on shallow, dry, infertile soils. Stressors include nonnative species and improper use by livestock.

Low and black sagebrush have a more limited distribution than other sagebrush species and depend on seeding for regeneration, so their distribution could be further restricted in a warmer climate, resulting in a patchier mosaic of remnant communities (West and Mooney 1972). Several traits make low sagebrush sensitive to climate change, including high mortality in the first year of growth (Shaw and Monsen 1990), which may be exacerbated if erosion increases from drought-induced reductions of plant cover. If unfavorable conditions for seeding persist following disturbance, low sagebrush may disappear from some sites, especially if annual grasses increase at the same time.

Increased fire will have negative consequences for low and black sagebrush, which are intolerant of fire and do not resprout. Fire return intervals vary considerably among communities dominated by low sagebrush. In the Greater Yellowstone Area subregion, vegetation modeling results indicate that the proportion of landscape burned will increase substantially, with the potential for fire to reach some low sagebrush communities (Sheehan et al. 2015). Increased fire activity will decrease the abundance of low sagebrush relative to other species, especially if nonnative annual grasses become more prevalent.

Relative to other sagebrush species, low and black sagebrush have limited adaptive capacity. Black sagebrush hybridizes with silver sagebrush, and sprouting is thought to be a heritable trait in crosses between non-sprouting and sprouting sagebrushes (McArthur 1994). However, silver sagebrush distributions are typically disjunct with those of low and black sagebrush, so acquisition of sprouting traits is unlikely. The relatively low productivity of low sagebrush sites may also limit adaptive capacity, especially if other risk factors are present.

6.4.4 Sprouting Sagebrush Species

Significant areas of threetip and silver sagebrush shrublands have been converted to agricultural lands. Those that remain are often used for livestock grazing because of the palatable herbaceous undergrowth. Rangelands with improper grazing typically have a large amount of bare ground, low vigor of native herbaceous species, and presence of nonnative plant species. Improper livestock grazing can cause loss of topsoil if vegetation cover and density decline and bare ground increases (Sheatch and Carlson 1998).

Both species can sprout from the root crown following top kill from fire (Bunting et al. 1987); silver sagebrush is a vigorous sprouter (Rupp et al. 1997), whereas threetip sagebrush is less vigorous (Bunting et al. 1987; Akinsoji 1988). Both species occur on mesic sites, where threetip sagebrush is often associated with mountain big sagebrush, and silver sagebrush occupies riparian benches or moist toe slopes. Although these species sprout, increased fire frequency and severity (particularly in threetip communities) may facilitate dominance by fire-adapted shrubs, herbs, and nonnative species. More spring and winter precipitation would promote establishment and early seed set in nonnative annual grasses, providing a competitive advantage over native perennial grasses (Bradley 2008) and creating fine fuels that can burn sagebrush and native grasses before they have matured and set seed (Chambers and Pellant 2008).

Historical fire return intervals for both species are relatively short, with threetip sagebrush cover returning to preburn levels 30–40 years after fire (Barrington et al. 1988). All three subspecies of silver sagebrush sprout after fire, and along with threetip, also occur on more mesic sites. If high-severity burns are more frequent in a warmer climate, they may not only cause mortality, but create unfavorable conditions for postfire regeneration (from sprouting or seed), and provide invasive species with a competitive edge. Understory composition may shift to more xeric grassland species (e.g., bluebunch wheatgrass, needle-and-thread) that are better adapted to drier conditions. In a warmer climate, both sagebrush species may persist in sites that retain sufficient moisture (e.g., higher elevation, north aspect, concavities).

6.4.5 *Wyoming Big Sagebrush and Basin Big Sagebrush*

Distribution of Wyoming big sagebrush is patchy in Montana and more evenly distributed in the Eastern Rockies and Grassland subregions. Stressors to both sagebrush communities include improper livestock grazing, native ungulate herbivory, and nonnative invasive plants. Loss of topsoil can occur if vegetation cover and density decline and bare ground increases (e.g., with ungulate impacts). Wyoming big sagebrush habitat coincides with oil and gas development, especially on the eastern edge of its distribution. The Grassland and Greater Yellowstone Area subregions contain the largest extent of big sagebrush, with basin big sagebrush dominant in the Western Rockies subregion.

Big sagebrush ecosystems have been subjected to many stressors: oil and gas development (Knick et al. 2003), big sagebrush removal to increase livestock forage, plant pathogens and insects, improper grazing (Davies et al. 2011), nonnative invasive species (Davies et al. 2011), and altered disturbance regimes (Balch et al. 2013). These stressors, especially oil and gas development, cause habitat loss and fragmentation (Doherty et al. 2008; Walston et al. 2009), creating barriers to plant dispersal and degrading habitat for sage-grouse and other wildlife species (Rowland et al. 2006). Improper use by livestock alters structure and composition of big sagebrush communities and increases the probability of nonnative annual grass invasion (Cooper et al. 2007; Davies et al. 2011), especially cheatgrass (Brooks et al. 2004; Balch et al. 2013). Cheatgrass invasion poses a continued and heightened threat to big sagebrush ecosystems in the future, because its biomass production and fire frequency are projected to increase in response to rising temperature and CO_2 levels (Ziska et al. 2005). Field brome (*Bromus arvensis*) can also negatively affect big sagebrush, because it establishes after fires that kill big sagebrush (Cooper et al. 2007).

Amount and timing of precipitation control seedling establishment of big sagebrush at low elevation, whereas minimum temperature and snow depth control germination and survival at high elevation (Poore et al. 2009; Nelson et al. 2014). If drought increases in the future, germination and survival of big sagebrush seedlings may decrease (Schlaepfer et al. 2014a, b). Drought and increased summer temperature can also affect survival and growth of mature big sagebrush plants (Poore et al. 2009), as well as perennial grasses and forbs. In addition, big sagebrush seeds have low viability after 2 years (Wijayratne and Pyke 2009, 2012), are poorly dispersed (Young et al. 1989), and are episodically produced (Young et al. 1989). Big sagebrush is a poor competitor relative to associated herbaceous species (Schlaepfer et al. 2014a), and because it is killed by fire, postfire recovery may be challenging if wildfires are more frequent in the future.

Big sagebrush is projected to expand in northeast and north-central Montana, where climate may be sufficiently cool and moist (Schrag et al. 2011), and decrease in the Western Rockies and northwest Greater Yellowstone Area subregions, primarily from summer drought (Schlaepfer et al. 2012). Projected shifts in community composition and productivity in big sagebrush ecosystems remain uncertain. If drought increases, native herbaceous plant diversity and cover may be reduced. In non-drought years, higher temperatures and increased CO_2 may increase biomass production (Reeves et al. 2014), fire frequency, and herbaceous biomass at the expense of fire-intolerant big sagebrush.

Although lower soil water availability may pose a long-term stress for big sagebrush ecosystems, extended periods of sustained drought are required to cause mortality (Kolb and Sperry 1999). Big sagebrush should have some capacity to adapt to climate change. The species occurs over a large geographic area with diverse topography, soils, and climate, suggesting that it can persist in a broad range of ecological conditions. In addition, various subspecies of big sagebrush hybridize and have a high level of polyploidy, providing capacity to undergo selection and adapt to variable climate (e.g., Poore et al. 2009).

6.4.6 *Mountain Big Sagebrush*

Some areas of mountain big sagebrush shrublands have been converted to agricultural lands, and those that remain are used for domestic livestock grazing, primarily because of the palatable herbaceous undergrowth. Those that have had improper grazing typically have bare ground and low vigor of native herbaceous species, and as a result, nonnative plant species are often present. Improper livestock grazing, native ungulate herbivory, and nonnative invasive plants are the primary stressors. Fire exclusion is also a stressor, facilitating conifer establishment and decline of grass cover.

Mountain big sagebrush is killed by fire. If wildfire frequency and severity increase, community composition will shift to dominance by fire-adapted shrub, herbaceous, and nonnative species. Increased spring and winter precipitation may facilitate nonnative annual grasses (especially cheatgrass) establishment, although this is less likely in cooler locations compared to lower elevation Wyoming and basin big sagebrush. Concerns about cheatgrass and other nonnative species have been discussed above, although some sagebrush communities may be less susceptible to cheatgrass invasion following fire (Lavin et al. 2013).

Mountain big sagebrush is not fire adapted, and may decline in cover and density or become extirpated in response to warmer temperatures and increased fire frequency and severity. Historical fire return intervals in mountain big sagebrush were a few decades, compared to Wyoming big sagebrush (>100 years) (Heyerdahl et al. 2006; Lesica et al. 2007). Mountain big sagebrush regenerates from seeds, with full recovery 15–40 years after fire (Bunting et al. 1987). Because the sagebrush seed bank is minimal, if fires burn large areas and there are no live, seed-bearing sagebrush nearby, there may be a conversion to grassland. In addition, nonnative species may expand into these areas or increase in abundance (Bradley 2008).

Mountain big sagebrush occurs at higher elevations, typically on more productive cooler, mesic sites that are less susceptible to nonnative species. If these sites become warmer and drier, herbaceous understory composition could shift to more drought-tolerant species, including cheatgrass (Chambers et al. 2014). Mountain big sagebrush may be able to persist and expand into cooler locations (higher elevation, north aspects, concavities, deeper soils). Native grassland species that are more tolerant of warmer, drier conditions (e.g., bluebunch wheatgrass, needle-and-thread) may also be able to persist in the understory.

6.4.7 Northern Great Plains

Grasslands extend across the northern Great Plains, from the foothill grasslands along the east slope of the northern and central Rocky Mountains in Montana to the Red River basin in eastern North Dakota. Annual precipitation increases from west to east, with a concurrent transition from shortgrass prairie to northern mixed grass prairie to tallgrass prairie. Shortgrass prairie is characterized by grama (*Bouteloua* spp.)/needlegrass/wheatgrass and a mix of C3 and C4 plant species. Northern mixed grass prairie is characterized by wheatgrass/needlegrass in the west and wheatgrass/bluestem (*Andropogon* spp.)/needlegrass in the east, and a mix of C3 and C4 plant species. Tallgrass prairie is characterized by bluestem and a dominance of C4 grasses, although C3 grass species are also present.

Historically, frequent wildfire maintained grassland dominance, particularly in the eastern Great Plains. Starting in the late nineteenth century, settlement altered fire regimes by reducing fire frequency and changing the seasonality of fire. The predominant land use and land cover changed from grasslands to crop agriculture and domestic livestock production, affecting the continuity of fuels and fire spread. Reduced fire has encouraged woody plant encroachment, especially in the eastern Great Plains (Morgan et al. 2008). Invasive grass and forb species have reduced the diversity of native grasslands, with increased noxious weeds such as leafy spurge, Kentucky bluegrass (*Poa pratensis*), Japanese brome, and cheatgrass. Energy development and associated infrastructure fragments grassland locally, and roads and vehicles help spread nonnative species.

Soil water availability affects plant species distribution and abundance, productivity, and associated social and economic systems of the northern Great Plains. Interactions of temperature, precipitation, topography, soil, and ambient CO_2 with plant physiological mechanisms will influence how grasslands respond to climate change (Chen et al. 1996; Bachman et al. 2010; Morgan et al. 2011). Slope, aspect, insolation, and soil water holding capacity will modify these effects locally. Landscape variability in available soil water will result in uneven patterns of vegetation change and productivity. Elevated CO_2 may counter the effects of higher temperatures and evaporative demand by increasing water use efficiency of some plant species, especially C3 grasses (Morgan et al. 2011), although most nonnative invasive species are C3 plants, and expansion of nonnatives would be a negative outcome (Morgan et al. 2008).

The adaptive capacity of Great Plains grasslands was demonstrated in the Central Plains during the 1930s and 1950s droughts (Weaver 1968). There was a shift in C4 grasses, in which big bluestem (*Andropogon gerardii*) and little bluestem (*Schizachyrium scoparium*) were replaced by the shortgrass species blue grama (*Bouteloua gracili*) and buffalograss (*Bouteloua dactyloides*). Shifts from tallgrass prairie to mixed grass prairie were also documented with an increase in two C3 plants, western wheatgrass and needlegrass. This shift was later reversed during the higher precipitation period of the 1940s, indicating significant resilience of Great Plains grasslands to drought.

6.5 Adapting Rangeland Vegetation and Management to Climate Change

Rangeland vegetation in the northern Rockies will be affected by altered fire regimes, increased drought, and increased establishment of nonnative invasive species in a changing climate. Effects of climate change will compound existing stressors caused by human activities. Therefore, adaptation options for rangeland vegetation are focused on increasing the resilience of rangeland ecosystems, primarily through nonnative invasive species control and prevention.

Ecologically based invasive plant management (EBIPM) (Krueger-Mangold et al. 2006; Sheley et al. 2006) provides a framework for controlling nonnative species in rangelands. EBIPM focuses on strategies to repair damaged ecological processes that facilitate establishment of nonnatives (James et al. 2010). For example, prescribed fire treatments can be used where fire regimes have been altered, and seeding of desired natives can be done where seed availability and dispersal of natives is low.

Another adaptation strategy is to increase proactive management actions to prevent establishment of nonnative species. Early detection/rapid response (EDRR) is commonly used to prevent nonnative species establishment. Other tactics include implementing weed-free policies, conducting outreach to educate employees and the public about nonnatives (e.g., teach people to clean their boots), and developing weed management areas that are collaboratively managed by multiple agencies, non-governmental organizations, and the public.

Grazing management will be important in maintaining and increasing resilience of rangelands to climate change. A warmer climate will lead to altered availability of forage, requiring some reconsideration of grazing strategies. For example, reducing grazing in July and August may encourage growth of desired perennials in degraded systems. Livestock grazing can also be managed through development of site-specific, within-season triggers and end-point indicators that would inform livestock movement guides and allow for the maintenance and enhancement of plant health.

A changing climate has led to a decline of pollinators in some rangeland systems and may lead to phenological mismatches between pollinators and host plants. Pollinator declines may negatively affect the health of grasslands in the Northern Rockies, so encouraging native pollinators may help sustain these systems. Tactics that promote native pollinators include revegetation with native species, appropriate herbicide and insecticide use, and education. Implementing long-term monitoring of pollinators can help identify where treatments should be prioritized.

Existing stressors in montane shrublands include fire exclusion and conifer establishment, browsing by native and domestic ungulates, and insects and disease. Warmer temperatures and drier conditions may lead to an increase in high-severity fires that can cause extirpation of characteristic species and local soil erosion. Adaptation tactics include implementing fuel reduction projects such as brush cutting, slashing, mastication, and targeted browsing. Reestablishing appropriate fire regimes may help maintain these shrublands and increase their resilience to a warmer climate. EDRR and EBIPM can be used to control nonnatives and to maintain adequate shrub cover, vigor, and species richness. Educating specialists on ecology and disturbances affecting shrublands, effects of repeated burns, reforestation needs, and reporting on weeds will also help maintain these systems.

References

Akinsoji, A. (1988). Postfire vegetation dynamics in a sagebrush steppe in southeastern Idaho, USA. *Vegetatio, 78*, 151–155.

Allen, C. R., Gunderson, L., & Johnson, A. R. (2005). The use of discontinuities and functional groups to assess relative resilience in complex systems. *Ecosystems, 8*, 958–966.

Arno, S., & Gruell, G. (1986). Douglas-fir encroachment into mountain grasslands in southwestern Montana. *Journal of Range Management, 39*, 272–276.

Bachelet, D., Neilson, R. P., Lenihan, J. M., & Drapek, R. J. (2001). Climate change effects on vegetation distribution and carbon budget in the United States. *Ecosystems, 4*, 164–185.

Bachman, S., Heisler-White, J. L., Pendall, E., et al. (2010). Elevated carbon dioxide alters impacts of precipitation pulses on ecosystem photosynthesis and respiration in a semi-arid grassland. *Oecologia, 162*, 791–802.

Balch, J. K., Bradley, B. A., D'Antonio, C. M., & Gómez-Dans, J. (2013). Introduced annual grass increases regional fire activity across the arid western USA (1980–2009). *Global Change Biology, 19*, 173–183.

Barrington, M., Bunting, S., & Wright, G. (1988). *A fire management plan for Craters of the Moon National Monument, Cooperative Agreement CA-9000-8-0005*. Moscow: University of Idaho, Range Resources Department.

Boyte, S. P., Wylie, B. K., & Major, D. J. (2016). Cheatgrass percent cover change: Comparing recent estimates to climate change driven predictions in the Northern Great Basin. *Rangeland Ecology and Management, 69*, 265–279.

Bradford, J., Schlaepfer, D., & Lauenroth, W. (2014). Ecohydrology of adjacent sagebrush and lodgepole pine ecosystems: The consequences of climate change and disturbance. *Ecosystems, 17*, 590–605.

Bradley, B. A. (2008). Regional analysis of the impacts of climate change on cheatgrass invasion shows potential risk and opportunity. *Global Change Biology, 14*, 1–13.

Bradley, B. A., Curtis, C. A., & Chambers, J. C. (2016). *Bromus* response to climate and projected changes with climate change. In M. J. Germino (Ed.), *Exotic brome-grasses in arid and semi-arid ecosystems of the Western US*. Cham: Springer International Publishing.

Brooks, M. L., D'Antonio, C. M., Richardson, D. M., et al. (2004). Effects of invasive alien plants on fire regimes. *Bio Science, 54*, 677–688.

Bunting, S. C., Kilgore, B. M., & Bushey, C. L. (1987). *Guidelines for prescribed burning sagebrush-grass rangelands in the northern Great Basin, General Technical Report INT-231*. Ogden: U.S. Forest Service, Intermountain Research Station.

Chambers, J. C., & Pellant, M. (2008). Climate change impacts on northwestern and intermountain United States rangelands. *Rangelands, 30*, 29–33.

Chambers, J. C., Roundy, B. A., Blank, R. R., et al. (2007). What makes Great Basin sagebrush ecosystems invasible to *Bromus tectorum*? *Ecological Monographs, 77*, 117–145.

Chambers, J. C., Bradley, B. A., Brown, C. A., et al. (2014). Resilience to stress and disturbance, and resistance to *Bromus tectorum* L. invasion in the cold desert shrublands of western North America. *Ecosystems, 17*, 360–375.

Chen, D., Hunt, H. W., & Morgan, J. A. (1996). Responses of a C3 and C4 perennial grass to CO_2 enrichment and climate change: Comparison between model predictions and experimental data. *Ecological Modelling, 87*, 11–27.

Cooper, S. V., Lesica, P., & Kudray, G. M. (2007). *Postfire recovery of Wyoming big sagebrush shrub-steppe in central and southeast Montana*. Helena: Bureau of Land Management, State Office, Montana Natural Heritage Program.

Davies, K. W., Boyd, C. S., Beck, J. L., et al. (2011). Saving the sagebrush sea: An ecosystem conservation plan for big sagebrush plant communities. *Biological Conservation, 144*, 2573–2584.

Doherty, K. E., Naugle, D. E., Walker, B. L., & Graham, J. M. (2008). Greater sage-grouse winter habitat selection and energy development. *Journal of Wildlife Management, 72*, 187–195.

Folke, C., Carpenter, S., Walker, B., et al. (2004). Regime shifts, resilience, and biodiversity in ecosystem management. *Annual Review of Ecology, Evolution, and Systematics, 33*, 557–581.

Heyerdahl, E. K., Miller, R. F., & Parson, R. A. (2006). History of fire and Douglas-fir establishment in a savanna and sagebrush-grassland mosaic, southwestern Montana, USA. *Forest Ecology and Management, 230*, 107–118.

Holling, C. S. (1973). Resilience and stability in ecological systems. *Annual Review of Ecology and Systematics, 4*, 1–23.

James, J. J., Smith, B. S., Vasquez, E. A., & Sheley, R. L. (2010). Principles for ecologically based invasive plant management. *Invasive Plant Science and Management, 3*, 229–239.

Knick, S. T., Dobkin, D. S., Rotenberry, J. T., et al. (2003). Teetering on the edge or too late? Conservation and research issues for avifauna or sagebrush habitats. *The Condor, 105*, 611–634.

Kolb, K. J., & Sperry, J. S. (1999). Differences in drought adaptation between subspecies of sagebrush (*Artemisia tridentata*). *Ecology, 80*, 2373–2384.

Krueger-Mangold, J. M., Sheley, R. L., & Svejcar, T. J. (2006). Toward ecologically-based invasive plant management on rangeland. *Weed Science, 54*, 597–605.

Lavin, M., Brummer, T., Quire, J., et al. (2013). Physical disturbance shapes vascular plant diversity more profoundly than fire in the sagebrush steppe of southeastern Idaho, U.S.A. *Ecology and Evolution, 3*, 1626–1641.

Lesica, P., Cooper, S. V., & Kudray, G. (2007). Recovery of big sagebrush following fire in southwest Montana. *Rangeland Ecology and Management, 60*, 261–269.

Lund, G. H. (2007). Accounting for the worlds rangelands. *Rangelands, 29*, 3–10.

McArthur, E. D. (1994). Ecology, distribution, and values of sagebrush within the Intermountain region. In: S. B. Monsen, & S. G. Kitchen (Comps.), *Proceedings—Ecology and management of annual rangelands* (General Technical Report INT-GTR-313, pp. 347–351). Ogden: U.S. Forest Service, Intermountain Research Station.

Merrill, K. R., Meyer, S. E., & Coleman, C. E. (2012). Population genetic analysis of *Bromus tectorum* (Poaceae) indicates recent range expansion may be facilitated by specialist genotypes. *American Journal of Botany, 99*, 529–537.

Morgan, J. A., Mosier, A. R., Milchunas, D. G., et al. (2004a). CO_2 enhances productivity, alters species composition, and reduces digestibility of short grass steppe vegetation. *Ecological Applications, 14*, 208–219.

Morgan, J. A., Pataki, D. E., Körner, C., et al. (2004b). Water relations in grassland and desert ecosystems exposed to elevated atmospheric CO_2. *Oecologia, 140*, 11–25.

Morgan, J. A., Milchunas, D. G., LeCain, D. R., et al. (2007). Carbon dioxide enrichment alters plant community structure and accelerates shrub growth in the short grass steppe. *Proceedings of the National Academy of Sciences, USA, 104*, 14724–14729.

Morgan, J. A., Derner, J. D., Milchunas, D. G., & Pendall, E. (2008). Management implications of global change for Great Plains rangelands. *Rangelands, 30*, 18–22.

Morgan, J. A., LeCain, D. R., Pendall, E., et al. (2011). C4 grasses prosper as carbon dioxide eliminates desiccation in warmed semi-arid grassland. *Nature, 476*, 202–206.

Nelson, Z. J., Weisberg, P. J., & Kitchen, S. G. (2014). Influence of climate and environment on postfire recovery of mountain big sagebrush. *International Journal of Wildland Fire, 23*, 131–142.

Ortega, Y., Pearson, D. E., Waller, L. P., et al. (2012). Population-level compensation impedes biological control of an invasive forb and indirect release of a native grass. *Ecology, 93*, 783–792.

Palmquist, K. A., Schlaepfer, D. R., Bradford, J. B., & Lauenroth, W. K. (2016a). Spatial and ecological variation in dryland ecohydrological responses to climate change: Implications for management. *Ecosphere, 7*, e01590.

Palmquist, K. A., Schlaepfer, D. R., Bradford, J. B., & Lauenroth, W. K. (2016b). Mid-latitude shrub steppe plant communities: Climate change consequences for soil water resources. *Ecology, 97*, 2342–2354.

Polley, H. W., Briske, D. D., Morgan, J. A., et al. (2013). Climate change and north American rangelands: Trends, projections, and implications. *Rangeland Ecology and Management, 66*, 493–511.

Poore, R. E., Lamanna, C. A., Ebersole, J. J., & Enquist, B. J. (2009). Controls on radial growth of mountain big sagebrush and implications for climate change. *Western North American Naturalist, 69*, 556–562.

Ramakrishnan, A. P., Meyer, S. E., Fairbanks, D. J., & Coleman, C. E. (2006). Ecological significance of microsatellite variation in western north American populations of *Bromus tectorum*. *Plant Species Biology, 21*, 61–73.

Reeves, M. C., & Baggett, L. S. (2014). A remote sensing protocol for identifying rangelands with degraded productive capacity. *Ecological Indicators, 43*, 172–182.

Reeves, M. C., & Mitchell, J. E. (2011). Extent of coterminous U.S. rangelands: Quantifying implications of differing agency perspectives. *Rangeland Ecology and Management, 64*, 1–12.

Reeves, M., Moreno, A., Bagne, K., & Running, S. W. (2014). Estimating the effects of climate change on net primary production of US rangelands. *Climatic Change, 126*, 429–442.

Rowland, M. M., Wisdom, M. J., Spring, L. H., & Meinke, C. W. (2006). Greater sage-grouse as an umbrella species for sagebrush-associated vertebrates. *Biological Conservation, 129*, 323–335.

Rupp, L., Roger, K., Jerrian, E., & William, V. (1997). Shearing and growth of five intermountain native shrub species. *Journal of Environmental Horticulture, 15*, 123–125.

Schlaepfer, D. R., Lauenroth, W. K., & Bradford, J. B. (2012). Effects of ecohydrological variables on current and future ranges, local suitability patterns, and model accuracy in big sagebrush. *Ecography, 35*, 374–384.

Schlaepfer, D. R., Lauenroth, W. K., & Bradford, J. B. (2014a). Modeling regeneration responses of big sagebrush (*Artemisia tridentata*) to abiotic conditions. *Ecological Modelling, 286*, 66–77.

Schlaepfer, D. R., Lauenroth, W. K., & Bradford, J. B. (2014b). Natural regeneration processes in big sagebrush (*Artemisia tridentata*). *Rangeland Ecology and Management, 67*, 344–357.

Schrag, A., Konrad, S., Miller, B., et al. (2011). Climate-change impacts on sagebrush habitat and West Nile virus transmission risk and conservation implications for greater sage-grouse. *Geo Journal, 76*, 561–575.

Shaw, N. L., & Monsen, S. B. (1990). Use of sagebrush for improvement of wildlife habitat. In H. G. Fisser (Ed.), *Wyoming shrublands: aspen, sagebrush and wildlife management. Proceedings, 17th Wyoming shrub ecology workshop* (pp. 19–35). Laramie: University of Wyoming, Department of Range Management.

Sheatch, G. W., & Carlson, W. T. (1998). Impact of cattle treading on hill land. 1. Soil damage patterns and pasture status. *New Zealand Journal of Agricultural Research, 41*, 271–278.

Sheehan, T., Bachelet, D., & Ferschweiler, K. (2015). Projected major fire and vegetation changes in the Pacific Northwest of the conterminous United States under selected CMIP5 climate futures. *Ecological Modelling, 317*, 16–29.

Sheley, R. L., Mangold, J. M., & Anderson, J. L. (2006). Potential for successional theory to guide restoration of invasive-plant-dominated rangeland. *Ecological Monographs, 76*, 365–379.

Walston, L. J., Cantwell, B. L., & Krummel, J. R. (2009). Quantifying spatiotemporal changes in a sagebrush ecosystem in relation to energy development. *Ecography, 32*, 943–952.

Weaver, J. E. (1968). *Prairie plants and their environment: A fifty-year study in the Midwest.* Lincoln: University of Nebraska Press.

West, M., & Mooney, H. A. (1972). Photosynthetic characteristics of three species of sagebrush as related to their distribution patterns in the White Mountains of California. *American Midland Naturalist, 88*, 479–484.

Wijayratne, U. C., & Pyke, D. A. (2009). *Investigating seed longevity of big sagebrush (Artemisia tridentata), Open-file Report 2009-1146*. Reston: U.S. Geological Survey.

Wijayratne, U. C., & Pyke, D. A. (2012). Burial increases seed longevity of two *Artemisia tridentata* (Asteraceae) subspecies. *American Journal of Botany, 99*, 438–447.

Woodward, F. I., & Kelly, C. K. (2008). Responses of global plant diversity capacity to changes in carbon dioxide concentration and climate. *Ecological Letters, 11*, 1229–1237.

Young, J. A., Evans, R. A., & Palmquist, D. E. (1989). Big sagebrush (*Artemisia tridentata*) seed production. *Weed Science, 37*, 47–53.

Ziska, L. H., Reeves, J. B., & Blank, B. (2005). The impact of recent increases in atmospheric CO_2 on biomass production and vegetative retention of Cheatgrass (Bromus tectorum): Implications for fire disturbance. *Global Change Biology, 11*, 1325–1332.

Chapter 7
Effects of Climate Change on Ecological Disturbance in the Northern Rockies

Rachel A. Loehman, Barbara J. Bentz, Gregg A. DeNitto, Robert E. Keane, Mary E. Manning, Jacob P. Duncan, Joel M. Egan, Marcus B. Jackson, Sandra Kegley, I. Blakey Lockman, Dean E. Pearson, James A. Powell, Steve Shelly, Brytten E. Steed, and Paul J. Zambino

Abstract Disturbances alter ecosystem, community, or population structures and change elements of the biological and/or physical environment. Climate changes can alter the timing, magnitude, frequency, and duration of disturbance events, as well as the interactions of disturbances on a landscape, and climate change may already be affecting disturbance events and regimes. Interactions among disturbance regimes, such as the co-occurrence in space and time of bark beetle outbreaks and wildfires, can result in highly visible, rapidly occurring, and persistent changes in landscape composition and structure. Understanding how altered disturbance patterns and multiple disturbance interactions might result in novel and emergent landscape behaviors is critical for addressing climate change impacts and for designing

---

R.A. Loehman (✉)
U.S. Geological Survey, Alaska Science Center, Anchorage, AK, USA
e-mail: rloehman@usgs.gov

B.J. Bentz
U.S. Forest Service, Rocky Mountain Research Station, Logan, UT, USA
e-mail: bbentz@fs.fed.us

G.A. DeNitto • M.E. Manning • J.M. Egan • M.B. Jackson • S. Kegley • S. Shelly
B.E. Steed • P.J. Zambino
U.S. Forest Service, Northern Region, Missoula, MT, USA
e-mail: gdenitto@fs.fed.us; mmanning@fs.fed.us; jegan@fs.fed.us; mbjackson@fs.fed.us; skegley@fs.fed.us; sshelly@fs.fed.us; bsteed@fs.fed.us; pzambino@fs.fed.us

R.E. Keane • D.E. Pearson
U.S. Forest Service, Rocky Mountain Research Station, Missoula, MT, USA
e-mail: rkeane@fs.fed.us; dpearson@fs.fed.us

J.P. Duncan • J.A. Powell
Department of Mathematics and Statistics, Utah State University, Logan, UT, USA
e-mail: jacob.duncan@aggiemail.usu.edu; jim.powell@usu.edu

I.B. Lockman
U.S. Forest Service, Pacific Northwest Region, Portland, OR, USA
e-mail: blockman@fs.fed.us

land management strategies that are appropriate for future climates. This chapter describes the ecology of important disturbance regimes in the Northern Rockies region, and potential shifts in these regimes as a consequence of observed and projected climate change. We summarize five disturbance types present in the Northern Rockies that are sensitive to a changing climate—wildfires, bark beetles, white pine blister rust (*Cronartium ribicola*), other forest diseases, and nonnative plant invasions—and provide information that can help managers anticipate how, when, where, and why climate changes may alter the characteristics of disturbance regimes.

Keywords Disturbance • Wildfire • Beetles • Pathogens • Climate change • Resilience • Rocky Mountains

7.1 Introduction

The term *disturbance regime* describes the general temporal and spatial characteristics of a *disturbance agent*, such as insects, disease, fire, and human activity, and the effects of that agent on the landscape (Table 7.1). More specifically, a disturbance regime is the cumulative effect of multiple disturbance events over space and time (Keane 2013). Disturbances alter ecosystem, community, or population structures and change elements of the biological and/or physical environment (White and Pickett 1985). The resulting shifting mosaic of diverse ecological patterns and structures affects future patterns of disturbance, in a reciprocal, linked relationship that shapes the fundamental character of landscapes and ecosystems.

Climate changes can alter the timing, magnitude, frequency, and duration of disturbance events, as well as the interactions of disturbances on a landscape, and climate change may already be affecting disturbance events and regimes (Dale et al. 2001). Interactions among disturbance regimes, such as the co-occurrence in space and time of bark beetle outbreaks and wildfires, can result in highly visible, rapidly occurring, and persistent changes in landscape composition and structure. Understanding how altered disturbance patterns and multiple disturbance interactions might result in novel and emergent landscape behaviors is critical for addressing climate change impacts and for designing land management strategies that are appropriate for future climates (Keane et al. 2015a).

This chapter describes the ecology of important disturbance regimes in the Northern Rockies region, and potential shifts in these regimes as a consequence of observed and projected climate change. We summarize five disturbance types in the Northern Rockies that are sensitive to a changing climate—wildfires, bark beetles, white pine blister rust (*Cronartium ribicola*), other forest diseases, and nonnative plant invasions—and provide information that can help managers anticipate how, when, where, and why climate change may alter the characteristics of disturbance regimes.

7 Effects of Climate Change on Ecological Disturbance in the Northern Rockies

Table 7.1 Characteristics used to describe disturbance regimes

Disturbance characteristic	Description	Example
Agent	Factor causing the disturbance	Mountain pine beetle is the agent that kills trees
Source, cause	Origin of the agent	Lightning is a source for wildland fire
Frequency	How often the disturbance occurs or its return time	Years since last fire or beetle outbreak (scale dependent)
Intensity	A description of the magnitude of the disturbance agent	Mountain pine beetle population levels; wildland fire heat output
Severity	The level of impact of the disturbance on the environment	Percent mountain pine beetle tree mortality; fuel consumption in wildland fires
Size	Spatial extent of the disturbance	Mountain pine beetles can kill trees in small patches or across entire landscapes
Pattern	Patch size distribution of disturbance effects; spatial heterogeneity of disturbance effects	Fire can burn large regions but weather and fuels can influence fire intensity and therefore the patchwork of tree mortality
Seasonality	Time of year at which a disturbance occurs	Species phenology can influence wildland fires effects; spring burns can be more damaging to growing plants than autumn burns on dormant plants
Duration	Length of time of that disturbances occur	Mountain pine beetle outbreaks usually last for 3–8 years; fires can burn for a day or for an entire summer
Interactions	Disturbances interact with each other, climate, vegetation and other landscape characteristics	Mountain pine beetles can create fuel complexes that facilitate or exclude wildland fire
Variability	The spatial and temporal variability of the above factors	Highly variable weather and mountain pine beetle mortality can cause variable burn conditions resulting in patchy burns of small to large sizes

From Keane (2013)

7.2 Wildfire

7.2.1 Overview

Wildland fire was historically the most important and extensive landscape disturbance in the Northern Rockies region (Hejl et al. 1995). Wildfire emerged as a dominant process in North America after the end of the last glacial period, about 16,500–13,000 years before present, with rapid climate changes and increased tree cover (Marlon et al. 2009). In the Northern Rockies, many forest types are fire prone and fire adapted, meaning that fire is an integral and predictable part of their maintenance and ecological functioning.

The role of fire in ecosystems and its interactions with dominant vegetation is termed a *fire regime* (Agee 1993). Fire regimes are defined by fire frequency (mean number of fires per time period), extent, intensity (measure of the heat energy released), severity (net ecological effect), and seasonal timing (Table 7.2).

Table 7.2 Risk assessment for fire regime changes, based on expert opinion and information from literature summarized in this chapter

Fire regime component	Predicted direction of change	Main driver(s) of change	Projected duration of change	Likelihood of change
Ignitions	Unknown	Changes in lightning frequency and human-caused ignitions	Unknown	Unknown
Area burned	Increase	Increased fire season length, decreased fuel moistures, increased extreme fire conditions	Until a sufficient proportion of the landscape has been exposed to fire, thus decreasing fuel loads and increasing structural and species heterogeneity	High
Fire frequency	Increase	Increased ignitions, increased fuel loads, decreased fuel moistures, increased fire season length	In forested systems until a sufficient proportion of the landscape has been exposed to fire, reducing fuel loads and continuity; in grass and shrubland systems, until global climate stabilizes	Moderate
Average fire size	Increase	Increased fire season length, decreased fuel moistures, increased extreme fire conditions	Until a sufficient proportion of the landscape has been exposed to fire, thus increasing the likelihood that previous fires will restrict growth of current year fires	High
Fire season length	Increase	Increased temperatures, decreased precipitation, decreased winter snowpack, decreased runoff	Until the global climate system stabilizes; predicted to increase as climate changes become more severe	High
Fire severity	Increase	Decreased fuel moistures, increased extreme fire conditions	In dry forests, until fires decrease surface fuel loads; in mesic forests, if increased fire frequency decreases fuel loads	Moderate

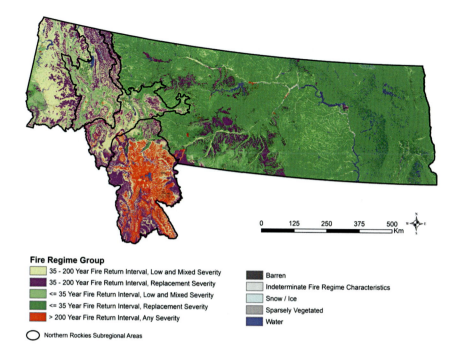

Fig. 7.1 Fire regime groups for the Northern Rockies, LANDFIRE mapping program. The fire regime group layer characterizes the presumed historical fire regimes within landscapes based on interactions among vegetation dynamics, fire spread, fire effects, and spatial context

Ecosystems in the Northern Rockies have been subject to a range of historical fire regimes, including (1) frequent (1–35 years), low- or mixed-severity fires that replaced less than 25% of the dominant overstory vegetation; (2) moderate-frequency (35–200 years), mixed-severity fires that replaced up to 75% of the overstory; and (3) infrequent (200+ years), high-severity fires that replaced greater than 75% of the dominant overstory vegetation (Fig. 7.1).

In general, fire regimes vary along environmental gradients, with fire frequency decreasing and fire severity increasing with elevation (although aspect and slope position can influence fire patterns). For example, low-severity fires are typical in many ponderosa pine (*Pinus ponderosa*) forests at low elevations. Historically, fires in ponderosa pine forests burned frequently enough to maintain low fuel loads and open stand structure, resulting in a landscape in which fire-caused mortality of mature trees was rare (Agee 1998; Jenkins et al. 2011). Conversely, high-severity fires occurring at intervals of more than 300 years are typical in subalpine forests. These fires cause extensive mortality of mature trees because long intervals between fires result in dense, multi-layer forest structures that are susceptible to crown fires (Agee 1998).

Climate and fuels are the two most important factors controlling fire regimes in forest ecosystems. Climate controls the frequency of weather conditions that promote fire, whereas the amount and arrangement of fuels influence fire intensity and spread. Climate influences fuels on longer time scales by shaping species composition and productivity (Dale et al. 2001; Marlon et al. 2008) and large-scale climatic patterns such as the El Niño Southern Oscillation and Pacific Decadal Oscillation are important drivers of forest productivity and susceptibility to disturbance in the Northern Rockies (Collins et al. 2006; Kitzberger et al. 2007). Current and past land use, including timber harvest, forest clearing, fire suppression, and fire exclusion through grazing, also affect the amount and structure of fuels (Falk et al. 2011).

At annual time scales, weather is the best predictor of fire characteristics such as area burned and fire size. In forest ecosystems, fuels lose moisture and become flammable in warm and dry summers typical in the Northern Rockies, during which time there are ample sources of ignition from lightning strikes and humans. Therefore, the active fire season (period conducive to active burning) is in the summer, typically from late June through October, with shorter seasons at higher elevation sites where snowpack can persist into July (Littell et al. 2009). In these high-elevation systems short-duration drying episodes generally do not create sufficiently dry conditions to sustain a fire, but prolonged dry weather conditions (about 40 days without precipitation) can sufficiently dry fuels to carry large, intense fires once they are ignited (Schoennagel et al. 2004). Regionally, widespread fire years are correlated with drought (Heyerdahl et al. 2008b; Morgan et al. 2008), and these regionally synchronous fires have generally occurred in the Northern Rockies (Idaho and western Montana) during years with relatively warm spring-summers and warm-dry summers (Heyerdahl et al. 2008a; Morgan et al. 2008).

In non-forested systems in the eastern Northern Rockies, precipitation amount, at both short (weeks to months) (Littell et al. 2009) and long (decades to centuries) (Brown et al. 2005) time scales is the dominant control on fire. During the fire season, the amount and timing of precipitation largely determine availability and combustibility of fine fuels, and short periods of dry weather are sufficient to precondition these systems to burn (Westerling and Swetnam 2003; Gedalof et al. 2005). In contrast to the grasslands of the southwestern United States, antecedent precipitation has not been found to be a significant driver of large fires in the northern grasslands; rather, large fires are most strongly correlated with low precipitation, high temperatures, and summer drought (July through September) in the year of the fire (Littell et al. 2009).

Humans are also important drivers of wildfire via altered ignition patterns associated with land clearing and land cover change, agriculture, introduction of nonnative species, and fire management (fuel treatments and fire suppression/exclusion). Grazing and the introduction of nonnative species have altered ecological processes that affect fire, including fuel loading and continuity, forest composition and structure, nutrient cycling, soils, and hydrology (Swetnam et al. 1999; Marlon et al. 2009). For many sagebrush ecosystems of low to moderate productivity, fire intervals are 10–20 times shorter today than what is estimated for pre-twentieth century conditions (Chap. 6), because of the spread and dominance of the nonnative annual cheatgrass (*Bromus tectorum*). In contrast, many dry forests, shrublands, and grasslands in the region exist in a state of "fire deficit" as the result of fire exclusion, leading to less frequent wildfire, higher stand densities, higher fuel quantities, and higher fuel continuity. This has increased the potential for crown fires in forests that historically experienced low-severity fire regimes (Peterson et al. 2005) and in some forests that experienced mixed-severity regimes (Taylor and Skinner 2003).

7.2.2 Potential Future Wildfire Regimes and Wildfire Occurrence

The most visible and significant short-term effects of climate changes on forest ecosystems are likely to be caused by altered disturbances, often occurring with increased frequency and severity. Climate changes are likely to increase fire frequency, fire season length, and cumulative area burned in the coming decades in the western United States, in response to warmer, drier conditions (McKenzie et al. 2004; Flannigan et al. 2006). Climate changes may also increase the frequency or magnitude of extreme weather events that affect fire behavior (Lubchenco and Karl 2012). Although shifts in vegetation composition and distribution caused by climate alone may occur over decades or centuries, wildfires can temporarily or persistently reorganize landscapes over a period of days (Overpeck et al. 1990; Seidl et al. 2011).

Earlier snowmelt, higher summer temperatures, and longer fire seasons have resulted in increased wildfire activity compared to the mid-twentieth century, particularly in the Northern Rockies (Westerling et al. 2006). Potential mid to late twenty-first century climate-driven changes to regional fire regimes include longer fire seasons and increases in fire frequency, annual area burned, number of high fire danger days, and fire severity as compared with modern fire patterns (Brown et al. 2004; Westerling et al. 2006; Rocca et al. 2014). In particular, lengthening of the fire season will allow for more ignitions, greater likelihood of fire spread, and a longer burning duration (Riley and Loehman 2016). A longer burning window, combined with regionally dry fuels, will promote larger fires and increased annual area burned relative to modern recorded fire activity. Earlier onset of snowmelt will reduce fuel moistures during the fire season, making a larger portion of the landscape flammable for longer periods of time (McKenzie et al. 2004). This shift may be especially pronounced in middle to high elevation forested systems where fuels are abundant. However, in areas that are fuel limited, fires may become more infrequent where there is insufficient moisture for fine fuel accumulation (Littell et al. 2009).

The potential effects of climate change on wildfire area have been assessed using statistical and ecological process models for the western United States (McKenzie et al. 2004; Spracklen et al. 2009), Pacific Northwest (Littell et al. 2010), Northern Rockies (Loehman et al. 2011; Holsinger et al. 2014; Rocca et al. 2014), and the Greater Yellowstone Area (Westerling et al. 2011). For a mean temperature increase of 2 °C, the annual area burned by wildfires is expected to increase by a factor of 1.4–5 for most western states (McKenzie et al. 2004). The effects of future climate on fire severity (i.e., the proportion of overstory mortality) are less certain because severity is more sensitive than area burned to arrangement and availability of fuels. The trend for larger, more damaging fires in sagebrush ecosystems is expected to continue until aberrations in fuel conditions that drive fire are altered (Keane et al. 2008).

7.2.3 Potential Interactions Between Wildfire and Other Disturbances

Interactions between fire and other disturbance agents will likely be a driver of ecosystem change under changing climate. Drought and increased temperatures are key drivers of both wildland fires and bark beetle outbreaks. Multiple studies have cited changes in fire behavior resulting from bark beetle-caused mortality in pine forests (reviewed in Hicke et al. 2012), with increased fire intensity persisting for approximately 5 years after fire, depending on local conditions.

Climate change may be a causal factor in recent increases in annual area burned by wildfires (Littell et al. 2009) and area affected by bark beetle outbreaks (Bentz et al. 2010). Projections of warmer temperatures and increased drought stress suggest that the total area susceptible to or affected by beetle outbreaks and large or severe fires may increase in the coming decades (Williams et al. 2013). Acting independently or synchronously in space and time, wildland fires and bark beetle outbreaks can substantially influence forest structure, composition, and function; abruptly reorganize landscapes; and alter biogeochemical processes such as carbon cycling, water supply, and nutrient cycles (Edburg et al. 2012; Hansen 2014).

7.3 Bark Beetles

7.3.1 Overview

Bark beetles are an important forest disturbance agent in the Northern Rockies region. Bark beetles in the region feed in the phloem of living conifers and can have extreme population amplifications over short time periods. Larval feeding, in addition to colonization by beetle-introduced fungi, typically results in death of the tree. Bark beetles are relative specialists, feeding on a single tree species or several species within a single genus, and in the Northern Rockies, multiple tree species are affected by different bark beetle species.

Historically, pulses of bark beetle-caused tree mortality were extensive across the northern Rocky Mountain region. Recently, between 1999 and 2013, bark beetle-caused tree mortality in the Northern Rockies affected nearly 570,000 hectares each year. Mountain pine beetle (*Dendroctonus ponderosae*, hereafter referred to as MPB) caused the majority of tree mortality, cumulatively affecting about 3.5 million hectares over the 1999–2013 time period. Across western North America between 1997 and 2010, bark beetle-caused tree mortality resulted in a transfer of carbon that exceeded that of fire-caused tree mortality (Hicke et al. 2013).

Bark beetle disturbances play a significant role in successional pathways and biogeochemical cycles in Northern Rockies forests (DeRose and Long 2007; Edburg et al. 2012; Hansen 2014). At low population levels, bark beetles act locally as thinning agents, producing forest gaps that promote regeneration and the release and subsequent growth of neighboring host and non-host trees, often producing uneven-aged stands (Mitchell and Preisler 1998). At outbreak population levels, tree mortality can approach 80% across landscapes of homogeneous host species and age, changing age-class distributions and overstory and understory species compositions. For example, in seral lodgepole pine forests removal of the largest trees by MPB can hasten succession by climax species when fire is absent (Hagle et al. 2000). Bark beetle disturbance can have long-term effects on forest structure and composition (Pelz and Smith 2012), and future landscape patterns in some forest types will be driven by tree mortality caused by large outbreaks of beetles.

7.3.2 Drivers of Bark Beetle Outbreaks

Bark beetle population outbreaks require forests with extensive host trees of suitable size and age (Fettig et al. 2013). For most irruptive species, preferred hosts are large, mature trees that provide a large amount of phloem resource for a developing brood. Large landscapes of these mature stands provide the perfect scenario for years of bark beetle population growth.

Although suitable host trees are critical to outbreak development, beetle populations can exist for years at low levels until release is triggered by inciting factors that allow for rapid population growth. Triggers include factors that increase survival and reproduction of the beetles. Stand conditions (Fettig et al. 2013), drought (Chapman et al. 2012; Hart et al. 2013), and pathogens (Goheen and Hansen 1993) can make it easier for low levels of beetles to overwhelm and kill trees. Similarly, large areas of host trees recently killed by fire, wind, or avalanche provide pulses of accessible food, and have resulted in outbreaks of some species such as Douglas-fir beetle (*Dendroctonus pseudotsugae*) and spruce beetle (*D. rufipennis*) (Shore et al. 1999; Hebertson and Jenkins 2007), as well as secondary beetles including *Ips* species and fir engraver (*Scolytus ventralis*) (Livingston 1979). Weather favorable to beetle reproduction and survival also influences population fluctuations, and can both initiate and sustain outbreaks (Régnière and Bentz 2007; Powell and Bentz 2009).

Climate and weather directly drive bark beetle outbreaks by affecting beetle growth and survival. For example, the process of mass attack needed to successfully overcome tree defenses requires synchronous emergence of adults, a process mediated by temperature (Bentz et al. 1991). Diapause and development rate thresholds help in this synchrony (Hansen et al. 2001, 2011; Bentz and Jönsson 2015).

Temperature is also an important determinant of the number of bark beetle generations per year. The western pine beetle (*D. brevicomis*) and *Ips* species can be bivoltine (two generations in one year) in the Northern Rockies (Kegley et al. 1997), although multivoltine in more southern parts of their range. Other bark beetle species require at least one year to complete a generation (univoltine), and at higher elevations where temperatures are cooler, two to three years may be required for a complete life cycle. Warm temperatures in the summer and spring extend the time that temperatures are above development thresholds, thereby allowing a reduction in generation time (Hansen et al. 2001; Bentz et al. 2014). Shorter generation times can lead to increased population growth, causing increased tree mortality.

Winter temperature also influences bark beetle population success. Larvae cold-harden to survive subfreezing temperatures (Bentz and Mullins 1999). However, extreme fluctuations in temperature in spring and autumn, in addition to long durations of temperatures below −35 °C, can cause extensive larval mortality (Safranyik and Linton 1991; Régnière and Bentz 2007).

7.3.3 Potential Effects of Climate Change on Bark Beetles

Climate change will likely have direct and indirect effects on bark beetle population outbreaks (Table 7.3). Indirectly, changing temperature and precipitation regimes influence the suitability and spatial distribution of host trees. Fungi, predators, and competitors associated with beetles can also be affected by changing climate and thereby indirectly affect beetle population outbreaks. Direct effects may also occur as changing temperature regimes either promote or disrupt bark beetle temperature-dependent life history strategies. Future bark beetle-caused tree mortality will therefore depend not only on the spatial distribution of live host trees and heterogeneity

Table 7.3 Risk assessment for mountain pine beetle outbreaks, developed using model simulations and expert opinion and information from literature summarized in this chapter

Elevation	Direction of change	Main driver(s) of change	Projected duration of change	Likelihood of change
<1000 m	Increase if host trees available	Temperature–caused shift to bivoltinism[a]	Increasing risk through 2100	High
1000–2000 m	Decrease	Temperature-caused disruption of seasonality	Decreasing risk through 2100	High
2000–3000 m	Increase initially, then decrease	Initially temperature-caused shift from semivoltine[b] to univoltine[c], then disruption of seasonality	Decreasing risk through 2100	High
>3000 m	Increase	Temperature-caused shift from semivoltine to univoltine	Increasing risk through 2100	High

[a]Two generations in 1 year
[b]One generation in 2 years
[c]One generation in 1 year

of future landscapes, but also on the ability of beetle populations and their associates to adapt to changing conditions (Bentz et al. 2016).

Projected changes in temperature and precipitation may cause significant stress to bark beetle host trees in the future. For example, host tree defenses can be weakened by reduced water availability (Chapman et al. 2012; Gaylord et al. 2013; Hart et al. 2013). Increasing temperatures are expected to alter the seasonal timing of soil water availability because of reduced snowpack and more precipitation falling as rain rather than snow (Regonda et al. 2005). Reduced soil water availability during the late spring and summer will lead to increased physiological drought stress in host trees that could indirectly benefit bark beetles that colonize stressed hosts in the late spring or summer (Raffa et al. 2008). Similarly, increased disturbance events could provide a reservoir of stressed trees used by some bark beetle species, leading to epidemic population levels.

Warming temperatures will also directly influence bark beetle population success, although the effects will depend on the beetle species, as well as the seasonal timing, amount, and variability of thermal input. For example, an increase in minimum temperature between 1960 and 2011 was associated with an increase in MPB survival and subsequent beetle-caused tree mortality in the Northern Rockies (Weed et al. 2015). As climate continues to change, extreme within-year variability in winter warming could be detrimental to insect survival, so reduced snow levels could therefore contribute to increased mortality.

Warming at other times of the year could similarly have both positive and negative effects on bark beetle populations. Phenological flexibility allows some species to shift voltinism pathways, developing on a semivoltine (one generation every 2 years) life cycle in cool years, and a univoltine lifecycle in warm years (Hansen et al. 2001; Bentz et al. 2014). Warming temperatures could also cause species that are currently bivoltine (e.g., western pine beetle, *Ips* species) to become multivoltine. These types of voltinism shifts can lead to rapid increases in beetle populations and subsequent tree mortality. Some thermal regimes allow these life cycle shifts yet maintain seasonal flights. However, other thermal regimes that result in voltinism shifts could disrupt seasonality (Régnière et al. 2015).

7.3.4 Projected Effects of Climate Change on Bark Beetle Populations

Mechanistic models can be used to explore the potential effects of changing climate on bark beetle populations. Here we describe results from a temperature-dependent mechanistic demographic model of MPB population growth that is based on phenological synchrony (Powell and Bentz 2009). The model was driven with downscaled temperatures from two GCMs (CanEMS2, CCSM4) and two greenhouse gas emissions scenarios (RCP 4.5 and 8.5). Climate data were downscaled using the multivariate adaptive constructed analogs approach (Abatzoglou and Brown 2012). Although indirect effects of climate clearly affect host tree vigor, stand composition, and distribution across a landscape, these effects were not included in the

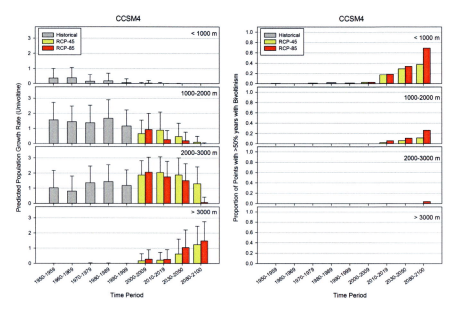

Fig. 7.2 Left panel: projected mountain pine beetle population growth rate (mean, standard deviation) of univoltine populations (one generation per year) over decades (historical) and 20-year periods (projected) from 1950 to 2100, for the RCP 4.5 and 8.5 emission scenarios (see section 7.3.4). Shown are the mean and standard deviation among locations of decadal (historical) and 2-decadal (projected) growth rates. Right panel: proportion of simulation points in which bivoltinism (two generations in 1 year) is projected for more than 50% of years in each time period

demographic model. Model output was considered only for locations where pines currently grow.

The proportion of areas with thermal requirements for MPB bivoltinism has historically been low in the Northern Rockies (Fig. 7.2). Stands at elevations <1000 meters currently have relatively few pines and low hazard for MPB, and population growth of univoltine populations was historically very low. This is most likely because it was too warm, and adult emergence synchrony was disrupted. Growth rate is projected to decrease further in current and future climates relative to historical periods. However, the proportion of simulation points at <1000 meters with thermal regimes that allow for bivoltinism is projected to increase through 2100, particularly with greater temperature increases (under RCP 8.5) (Fig. 7.2). The availability of pines at <1000 meters in future climates may be restricted.

Overall, model results suggest that pine stands above 2000 meters, particularly between 2000 and 3000 meters, have the highest risk to MPB-caused tree mortality in the near future. The highest density of pine occurs at 2000–3000 meters, the elevation range also associated with a majority of stands with high hazard (56%). These stands are projected to have higher univoltine population growth rates than in the historical period. Thermal regimes for bivoltinism are unlikely at these elevations (Fig. 7.2). In stands above 3000 meters, population growth rates were historically very low until 2000–2009. However, rates are projected to increase through 2100 (Fig. 7.2). These stands have historically been, and will remain, too cool for bivoltinism.

7.4 White Pine Blister Rust

7.4.1 Overview

White pine blister rust (*Cronartium ribicola,* hereafter referred to as WPBR) is a nonnative fungus introduced to western North America from Europe around 1910 (Tomback and Achuff 2011). The life cycle of WBPR requires two hosts, with two spore-producing stages on white pine and three separate spore producing stages on alternate hosts: *Ribes*, *Pedicularis,* and *Castilleja*. The WPBR fungus infects three white pines found in the Northern Region: western white pine (*Pinus monticola*), whitebark pine (*P. albicaulis*), and limber pine (*P. flexilis*). WPBR has been found across most of the ranges of these three pines, causing over 90% mortality in western white pine. WPBR infections rates are highest in the warmer, moister parts of the ranges of whitebark and limber pine (Tomback and Achuff 2011).

The time required for WPBR to kill its host varies by species, distance of infection from bole, and bole circumference. WPBR kills western white pine in 5–10 years, and whitebark pine in 20 years (Hoff and Hagle 1990). WPBR-caused tree mortality greatly affects stand structure and species composition, but the most serious impact of WPBR is the long-term impact on white pine regeneration. Native pine populations show some heritable resistance to WPBR, but the frequency of resistance is low and variable (Hoff et al. 1980), although resistance may have increased since this early report through additional natural selection (Klopfenstein et al. 2009).

7.4.2 Effects of Climate Change on WPBR

Climate change may cause WPBR infections to occur earlier and with greater frequency and intensity in pine stands. Specific weather conditions required for basidiospore germination and infection of pine needles may occur more frequently and for longer periods in the future (Koteen 1999). "Wave" years—hot and humid weather conditions throughout most of the growing season that facilitate infections on pine and alternate hosts, followed by moist but cooler weather events for teliospore and basidiospore production and pine infection—may increase in the future for whitebark pine, although wave years may actually decrease for most temperate pine forests because of hotter, drier conditions in a changing climate (Sturrock et al. 2011). Warmer temperatures could negatively impact rusts, although extreme weather could facilitate WPBR spore dispersal, resulting higher spore loads and expansion of its range (Helfer 2014).

Climate-mediated changes in host regeneration dynamics could restrict or expand host ranges (Helfer 2014), thus altering WPBR range. The distribution and occurrence of synergistic combinations of alternate host species (Zambino 2010) could also change. Higher elevation areas may experience new climates that facilitate the expansion of *Ribes* into areas that were historically too cold and snowy. On the other hand, low-elevation upland areas where *Ribes* are currently abundant might experience drought that causes decline of the host. Moreover, drought may cause extended and extensive stomatal closure in pines, thus preventing hyphae entry.

7.4.3 Interactions with Other Disturbance Processes

Interactions of fungal pathogens and their hosts with other disturbances may be a key factor in future WPBR infections (Ayres and Lombardero 2000). The interactive effects of wildland fire on WPBR are probably most important, but they are mostly minor and primarily indirect under future climates. The exception is the possibility that smoke may kill rust spores produced at the time of the fire (Hoffman et al. 2013).

Fire indirectly affects WPBR by changing the size, distribution, and abundance of its hosts. Mixed- and high-severity fires are currently common in most forests where WPBR is present (Arno et al. 2000), and these fires are projected to increase in size, frequency, and intensity (Westerling et al. 2011). Increased fire frequency and area burned can create favorable conditions for pine regeneration, because most five-needle pine seeds are dispersed by rodents and birds and are thus well adapted to spread into postfire landscapes. *Ribes* populations may increase after fire through regeneration by seed and sprouting from roots and rhizomes. However, re-burns soon after an initial fire can eliminate regenerating *Ribes* before they can develop a seed bank for the next forest regeneration cycle (Zambino 2010). Severe fires that kill rust-resistant pine trees may ensure continued high rust mortality in the future because they dampen the rate of rust-resistant adaptations (Keane et al. 2012). However, where rust-resistant pines survive fire, they can provide seeds for populating future landscapes that are resilient to rust infection and fire mortality.

Trees infected with WPBR are weakened, and may be more susceptible to fire-caused damage and mortality (Stephens and Finney 2002), and canopies of trees attacked or killed by WPBR may increase crown fire because of excessive pitch. Mortality from WPBR often results in elimination or thinning of the shade-intolerant pine overstory, allowing shade-tolerant competitors to occupy the openings and creating different canopy fuel conditions (Reinhardt et al. 2010). Many shade-tolerant competitors are more susceptible to fire damage, resulting in higher postfire tree mortality in rust-infected landscapes.

Mountain pine beetle also influences WPBR through regulation of the tree species that host both disturbance agents and killing of host trees that are resistant to the rust (Campbell and Antos 2000). For example, although whitebark pine stands in the Greater Yellowstone Area show little WPBR-related mortality, levels of MPB-related mortality are high (Kendall and Keane 2001; Macfarlane et al. 2013). Many stands of healthy pines in Yellowstone have been subjected to a major MPB outbreak over the last decade, resulting in substantial mortality of rust-resistant whitebark pine trees (Logan et al. 2008).

Model simulations of MPB disturbance under current climate suggest a decline in both lodgepole pine (especially without fire) and whitebark pine, with a corresponding increase in subalpine fir (*Abies lasiocarpa*) and Douglas-fir (*Pseudotsuga menziesii*), and little change with the addition of WPBR (Fig. 7.3). These trends are enhanced under a warmer climate, in which lodgepole pine declines are greater and stands are mainly replaced by Douglas-fir, but WPBR interaction has minor effects on species composition (Keane et al. 2015a). Fire frequency under current climate is 10% lower when fire, MPB, and WPBR are allowed to interact, and average tree mortality is also lower (Fig. 7.3). In a warmer climate, fire frequency decreases, high-severity fires increase, and interactions among disturbances create different landscapes than when each disturbance acts separately (or in the absence of disturbance) (Keane et al. 2015a) (Fig. 7.4).

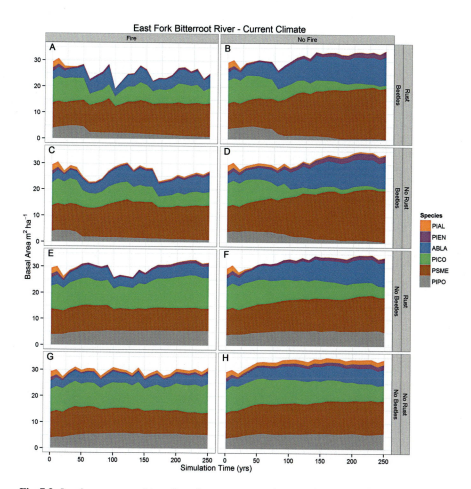

Fig. 7.3 Landscape composition of species cover types using the plurality of basal area for current climate for the East Fork of the Bitterroot River landscape with all combinations of fire, white pine blister rust, and mountain pine beetle. Species: PIAL = whitebark pine, PIEN = Engelmann spruce, ABLA = subalpine fir, PICO = lodgepole pine, PSME = Douglas-fir, and PIPO = ponderosa pine. Produced using the FireBGCv2 mechanistic ecosystem-fire process model (Keane et al. 2015a)

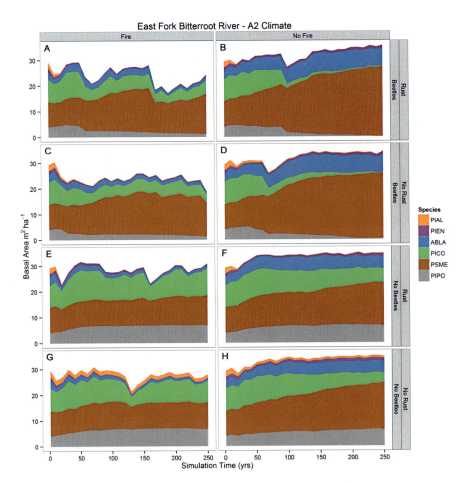

Fig. 7.4 Landscape composition of species cover types using the plurality of basal area for a warmer climate (A2 emission scenario) for the East Fork of the Bitterroot River landscape with all combinations of fire, white pine blister rust, and mountain pine beetle. Species: PIAL = whitebark pine, PIEN = Engelmann spruce, ABLA = subalpine fir, PICO = lodgepole pine, PSME = Douglas-fir, and PIPO = ponderosa pine. Produced using the FireBGCv2 mechanistic ecosystem-fire process model (Keane et al. 2015a)

7.5 Forest Diseases

7.5.1 Overview

We focus on forest diseases in the Northern Rockies known to have significant effects on ecosystems and ecosystem services, and for which at least some information on climate effects is available. These include dwarf mistletoes, root disease, needle disease, abiotic disease, and canker disease. Climate drivers and potential effects of climate change on these diseases are discussed in the following sections.

Dwarf mistletoes (*Arceuthobium* spp.) comprise five species of parasitic seed plants found in the Northern Rockies. They mainly cause reduced tree growth and productivity, but in some cases, they also cause tree mortality. Drummond (1982) estimated that 850,000 hectares of national forest lands were infested by the three most important species of dwarf mistletoe in the Northern Rockies.

Caused by various species of fungi, root disease is a major cause of tree growth loss and mortality in the Northern Rockies. The two most significant native pathogens in the Northern Rockies region are Armillaria (*Armillaria* spp.) and annosus (*Heterobasidion annosum*) root diseases, which occur in many mesic to moist forests west of the Continental Divide. At least 1.3 million hectares in the Northern Rockies have moderate to severe root disease, with up to 60% caused by *Armillaria ostoyae* (USFS 2007). Armillaria kills conifers of all species when they are young, but it is especially damaging to Douglas-fir, subalpine fir, and grand fir (*Abies grandis*) because these species remain susceptible throughout their lives (Kile et al. 1991). Armillaria and other root diseases influence forest species composition, structure, and successional trajectories by accelerating a transition to species that are more tolerant of root disease or by maintaining stands of more susceptible species in early seral stages (Byler and Hagle 2000).

Needle diseases have historically been of limited significance in the Northern Rockies; severe infection years occur only occasionally, and effects are mostly limited to crown thinning and loss of lower branches, with some mortality of young trees. However, periodic outbreaks can cause severe damage locally (Lockman and Hartless 2008). Needle diseases are favored by long, mild, damp springs, their occurrence at epidemic levels depending on favorable weather conditions and presence of an adequate host population.

Canker diseases, which affect tree branches and boles, typically occur in stressed trees that are poorly adapted to the sites in which they are growing. Damage is caused by breakage at the site of the cankers, or by mortality of branches and boles beyond girdling cankers.

Forests in the Northern Rockies periodically experience damage from weather extremes, such as high temperatures and drought. Conifers on the east side of the Continental Divide, primarily Douglas-fir and lodgepole pine, often experience twig and needle necrosis and desiccation caused by strong, dry, warm Chinook winds in winter (Bella and Navratil 1987). Drought injury can initiate a decline syndrome by predisposing trees to infection by less aggressive biotic agents, such as canker fungi and secondary beetles.

7.5.2 *Effects of Climatic Variability and Change on Forest Diseases*

Climatic variability and change can alter patterns of pathogen distribution and abundance through (1) direct effects on development and survival of a pathogen, (2) physiological changes in tree defenses, and (3) indirect effects on abundance of

Table 7.4 Risk assessment for forest diseases, developed using expert opinion and information from literature summarized in this chapter

Pathogen component	Direction of change	Main driver(s) of change	Projected duration of change	Likelihood of change
Needle disease	Significant increase if appropriate precipitation timing occurs	Increased precipitation in spring and early summer	May occur sporadically in association with weather events	High
Root disease	Little change	Host stress	While hosts are maladapted	Moderate
Dwarf mistletoe	Could decrease mistletoe populations	Temperature could influence flowering and seed production/dispersal	Unknown	Low
Abiotic disease	Significant increase	Temperature and decreased precipitation	Unknown	High

natural enemies, mutualists and competitors (Ayres and Lombardero 2000). Changing interactions between pathogens and their hosts may become a substantial driver of future disease outbreaks (Sturrock et al. 2011) (Table 7.4).

Fungi cause most forest diseases in the Northern Rockies, and fungus life cycles are directly influenced by climate-related factors such as timing and duration of precipitation, humidity, and temperature for spore germination, fungus growth, and inactivation. Dwarf mistletoe reproduction and infection are also affected by temperature and moisture (Hawksworth and Wiens 1996). Spring precipitation is projected to increase in most of the mountainous area of the Northern Rockies (Chap. 2). This may affect pathogens, specifically increasing the frequency of years when needle diseases cause significant needle loss in conifer species. For example, needle loss from Swiss needle cast (caused by *Phaeocryptopus gaeumannii*) is highly correlated with increasing winter temperatures and spring needle wetness (Stone et al. 2008).

Increased host stress in a changing climate could result in increased disease occurrence (Coakley et al. 1999). For example, low soil moisture (drought) affects the incidence and severity of pathogens. Flooding and water table fluctuation can also predispose trees to pathogens. Some diseases may become more damaging if thresholds that trigger infections, such as recurring drought, are reached more frequently.

Indirect effects of climate change on competitors, antagonists, and mutualists may also affect pathogens (Kliejunas et al. 2009). Some of the most profound effects of temperature and moisture changes could be on soil microflora, and on and in roots and shoots where a complex of organisms lives. Given that root pathogens of trees can often exploit a large food reserve in a tree once a defense is breached and use those reserves to bolster attacks on nearby trees, even small changes in relationships among fungal communities could have large effects.

Kliejunas (2011) conducted a qualitative risk assessment of the potential effects of climate change on forest diseases, several of which occur in the Northern Rockies. Risk potential for Dothistroma needle blight, (caused by *Dothistroma septosporum*) was estimated to be low in a warmer and drier climate, but moderate in a warmer and wetter climate. Risk potential for dwarf mistletoes was rated as high regardless of precipitation levels because dwarf mistletoe survival and infection increases with temperature. Armillaria root disease risk potential was rated as high to very high depending on moisture availability, with drier conditions increasing risk.

7.5.3 Forest Pathogen Interactions

Fire directly and indirectly influences distribution, severity, and persistence of forest diseases, and forest diseases can influence fire behavior and severity. Forest pathogens are directly damaged by smoke and heat of fires. Smoke can inhibit dwarf mistletoe seed germination (Zimmerman and Laven 1987), and heat from fire can kill pathogens that cause root disease in the upper soil profile (Filip and Yang-Erve 1997). High-intensity fires can completely remove a pathogen with its host (Kipfmueller and Baker 1998) or remove species susceptible to root disease (Hagle et al. 2000). In contrast, low-intensity fires can leave mosaics of pathogens along with their susceptible hosts, which can increase diseases such as dwarf mistletoe (Kipfmueller and Baker 1998). However, low-intensity fires in some habitats maintain species tolerant of root disease (e.g., western larch) (Hagle et al. 2000).

Human-caused fire exclusion has led to an increase of root disease and dwarf mistletoe (Hagle et al. 2000), which can influence fire behavior and severity. Root disease creates pockets of mortality, resulting in standing and downed woody debris and increased fuel loading. Increased litter accumulation and resinous witch's brooms from dwarf mistletoe infections can provide ladder fuels that may cause a ground fire to move into the canopy (Geils et al. 2002).

An increase in severe weather events and/or fires could increase occurrence of other diseases in a changing climate. For example, root and bole wounds could be used as "infection courts" for root disease, and such wounds from windfalls and fire are major avenues of infection for true fir and western hemlock (Smith 1989) and lodgepole pine (Littke and Gara 1986). Fire damage and other stresses can release root disease infections that have been walled off by host resistance responses (Hagle and Filip 2010).

Pathogens, insects, and fire can also interact. For example, root damage from fire in lodgepole pine can lead to stem decay fungi, which over time can cause extensive heartwood decay in the boles of trees (Littke and Gara 1986). Decay-infected trees then grow at a slower rate than uninfected trees and can be preferentially attacked by MPBs years later (Littke and Gara 1986). In addition, altered stand structure following MPB epidemics may increase dwarf mistletoe in lodgepole pine stands, thereby reducing stand growth and productivity and slowing stand recovery (Agne et al. 2014).

7.6 Nonnative Plants

7.6.1 Overview

Hundreds of nonnative species have been introduced into the Northern Rockies. Not all of these species are abundant, but recent surveys showed that nonnative plants comprise an average of 40% of species present (richness), and 25% of those nonnatives have significant effects on native grassland flora (Ortega and Pearson 2005). Most nonnative invasives are herbaceous species (graminoids and forbs), but some are shrub and tree species that commonly occur in riparian areas (e.g., Russian olive [*Elaeagnus angustifolia*], tamarisk [*Tamarix ramosissima*]). Nonnative invasive plant species represent a threat to ecosystem integrity, because they compete with native species in many plant communities and can alter ecological processes. These negative impacts can reduce biological diversity and forage for wildlife.

In this section we explore how climate change might alter current ecosystems and their susceptibility to invasion, and invasiveness of nonnative plants in general. We define the parameters that bound potential community change based on climate projections and discuss how plant communities might be affected across that range of conditions.

7.6.2 Effects of Climate Change on Nonnative Species

Numerous attributes associated with successful invaders suggest nonnatives could flourish under certain climate change scenarios (Dukes and Mooney 1999; Thuiller et al. 2008; but see Bradley et al. 2009). For example, many nonnatives are fast growing, early-seral species that tend to respond favorably to increased resource availability, including temperature, water, sunlight, and CO_2 (Walther et al. 2009). As a result, nonnatives often respond favorably to disturbance because disturbances can increase resource availability (Davis et al. 2000). Successful invaders commonly have strong dispersal strategies and shorter generation times, both of which can allow them to migrate more quickly than slow-growing and slowly-dispersed species (Clements and Ditommaso 2011). Greater plasticity of successful invaders could also favor their survival and ability to expand their populations (Clements and Ditomasso 2011).

Soil moisture often drives species-specific responses to elevated temperatures. For example, experimentally increasing temperatures in a Colorado meadow system resulted in increases in native upland shrubs, with big sagebrush (*Artemisia tridentata*) increasing in drier conditions and shrubby cinquefoil (*Dasiphora fruticosa*) in wetter conditions (Harte and Shaw 1995). Recent experimental work in western Montana showed that reduced precipitation can significantly impact spotted knapweed (*Centaurea melitensis*), whereas native bluebunch wheatgrass (*Pseudoroegneria spicata*) populations were unaffected by the same drought stress

(Ortega et al. 2012). This result is consistent with historical observations of spotted knapweed declines following drought conditions (Pearson and Fletcher 2008). In Wyoming sagebrush-steppe systems, bluebunch wheatgrass outperformed both cheatgrass and medusahead (*Taeniatherum caput-medusae*) in dry years, but the opposite was true in wet years (Mangla et al. 2011). Community-level studies in other grasslands have shown that drought periods can shift vegetation away from annual grasses and forbs and toward drought-tolerant native perennial grasses (Tilman and El Haddi 1992).

The most susceptible plant communities in the Northern Rockies have low vegetation cover, high bare ground, and unproductive soils; various nonnative plant species exploit these more open sites. As fires and other disturbances increase in intensity and frequency, invasive species may dominate some native plant communities, although numerous factors such as fire resistance of native species, propagule availability, and variation in burn severity can affect establishment (Zouhar et al. 2008). Invasive species are generally adaptable and capable of relatively rapid genetic change, which can enhance their ability to invade new areas in response to ecosystem modifications (Clements and Ditomasso 2011), including short-term disturbance or long-term stressors.

Note Any use of trade names is for descriptive purposes only and does not imply endorsement by the US Government.

References

Abatzoglou, J. T., & Brown, T. J. (2012). A comparison of statistical downscaling methods suited for wildfire applications. *International Journal of Climatology, 32*, 772–780.
Agee, J. (1993). *Fire ecology of Pacific Northwest forests*. Washington, DC: Island Press.
Agee, J. K. (1998). The landscape ecology of Western forest fire regimes. *Northwest Science, 72*, 24–34.
Agne, M. C., Shaw, D. C., Woolley, T. J., & Queijeiro-Bolaños, M. E. (2014). Effects of dwarf mistletoe on stand structure of lodgepole pine forests 21-28 years post-mountain pine beetle epidemic in central Oregon. *PloS One, 9*, e107532.
Arno, S. F., Parsons, D. J., & Keane, R. E. (2000). Mixed-severity fire regimes in the northern Rocky Mountains: Consequences of fire exclusion and options for the future. In *Wilderness science in a time of change conference. Volume 5: Wilderness ecosystems, threat, and management* (Proceedings RMRS-P-15-VOL 5, pp. 225–232). Fort Collins: U.S. Forest Service, Rocky Mountain Research Station.
Ayres, M. P., & Lombardero, M. J. (2000). Assessing the consequences of global change for forest disturbance from herbivores and pathogens. *Science of the Total Environment, 262*, 263–286.
Bella, I. E., & Navratil, S. (1987). Growth losses from winter drying (red belt damage) in lodgepole pine stands on the east slopes of the Rockies in Alberta. *Canadian Journal of Forest Research, 17*, 1289–1292.
Bentz, B. J., & Jönsson, A. M. (2015). Modeling bark beetle responses to climate change. In F. Vega & R. Hofstetter (Eds.), *Bark beetles: Biology and ecology of native and invasive species* (pp. 533–553). London: Academic.
Bentz, B. J., & Mullins, D. E. (1999). Ecology of mountain pine beetle (Coleoptera: Scolytidae) cold hardening in the Intermountain West. *Environmental Entomology, 28*, 577–587.

Bentz, B. J., Logan, J. A., & Amman, G. D. (1991). Temperature-dependent development of the mountain pine beetle (Coleoptera: Scolytidae) and simulation of its phenology. *Canadian Entomologist, 123,* 1083–1094.

Bentz, B., Régnière, J., Fettig, C., et al. (2010). Climate change and bark beetles of the western United States and Canada: Direct and indirect effects. *Bioscience, 60,* 602–613.

Bentz, B., Vandygriff, J., Jensen, C., et al. (2014). Mountain pine beetle voltinism and life history characteristics across latitudinal and elevational gradients in the western United States. *Forest Science, 60,* 434–449.

Bentz, B. J., Duncan, J. P., & Powell, J. A. (2016). Elevational shifts in thermal suitability for mountain pine beetle population growth in a changing climate. *Forestry, 89,* 271–283.

Bradley, B. A., Oppenheimer, M., & Wilcove, D. S. (2009). Climate change and plant invasions: Restoration opportunities ahead? *Global Change Biology, 15,* 1511–1521.

Brown, T. J., Hall, B. L., & Westerling, A. L. (2004). The impact of twenty-first century climate change on wildland fire danger in the western United States: An applications perspective. *Climatic Change, 62,* 365–388.

Brown, K. J., Clark, J. S., Grimm, E. C., et al. (2005). Fire cycles in North American interior grasslands and their relation to prairie drought. *Proceedings of the National Academy of Sciences, USA, 102,* 8865–8870.

Byler, J. W., & Hagle, S. K. (2000). *Succession functions of forest pathogens and insects: Ecosections M332a and M333d in northern Idaho and western Montana* (FHP Rep. 00–09). Missoula: U.S. Forest Service, Northern Region, State and Private Forestry, Forest Health Protection.

Campbell, E. M., & Antos, J. A. (2000). Distribution and severity of white pine blister rust and mountain pine beetle on whitebark pine in British Columbia. *Canadian Journal of Forest Resources, 30,* 1051–1059.

Chapman, T. B., Veblen, T. T., & Schoennagel, T. (2012). Spatiotemporal patterns of mountain pine beetle activity in the southern Rocky Mountains. *Ecology, 93,* 2175–2185.

Clements, D. R., & Ditomasso, A. (2011). Climate change and weed adaptation: Can evolution of invasive plants lead to greater range expansion than forecasted? *Weed Research, 51,* 227–240.

Coakley, S. M., Scherm, H., & Chakraborty, S. (1999). Climate change and plant disease management. *Annual Review of Phytopathology, 37,* 399–426.

Collins, B. M., Omi, P. N., & Chapman, P. L. (2006). Regional relationships between climate and wildfire-burned area in the Interior West, USA. *Canadian Journal of Forest Research, 36,* 699–709.

Dale, V. H., Joyce, L. A., McNulty, S., et al. (2001). Climate change and forest disturbances. *Bioscience, 51,* 723–734.

Davis, M. A., Grime, J. P., & Thompson, K. (2000). Fluctuating resources in plant communities: A general theory of invasibility. *Journal of Ecology, 88,* 528–534.

DeRose, R. J., & Long, J. N. (2007). Disturbance, structure, and composition: Spruce beetle and Engelmann spruce forests on the Markagunt Plateau, Utah. *Forest Ecology and Management, 244,* 16–23.

Drummond, D. B. (1982). *Timber loss estimates for the coniferous forests of the United States due to dwarf mistletoes* (Rep. 83–2). Fort Collins: U.S. Forest Service, Forest Pest Management.

Dukes, J. S., & Mooney, H. A. (1999). Does global change increase the success of biological invaders? *Trends in Ecology and Evolution, 14,* 135–139.

Edburg, S. L., Hicke, J. A., Brooks, P. D., et al. (2012). Cascading impacts of bark beetle-caused tree mortality on coupled biogeophysical and biogeochemical processes. *Frontiers in Ecology and the Environment, 10,* 416–424.

Falk, D. A., Heyerdahl, E. K., Brown, P. M., et al. (2011). Multi-scale controls of historical forest-fire regimes: New insights from fire-scar networks. *Frontiers in Ecology and the Environment, 9,* 446–454.

Fettig, C. J., Gibson, K. E., Munson, A. S., & Negron, J. F. (2013). Cultural practices for prevention and mitigation of mountain pine beetle infestations. *Forest Science, 60,* 450–463.

Filip, G. M., & Yang-Erve, L. (1997). Effects of prescribed burning on the viability of *Armillaria ostoyae* in mixed-conifer forest soils in the Blue Mountains of Oregon. *Northwest Science, 71,* 137–144.

Flannigan, M. D., Amiro, B. D., Logan, K. A., et al. (2006). Forest fires and climate change in the 21st century. *Mitigation and Adaptation Strategies for Global Change, 11,* 847–859.

Gaylord, M. L., Kolb, T. E., Pockman, W. T., et al. (2013). Drought predisposes pinon-juniper woodlands to insect attacks and mortality. *New Phytologist, 198,* 567–578.

Gedalof, Z., Peterson, D. L., & Mantua, N. J. (2005). Atmospheric, climatic, and ecological controls on extreme wildfire years in the northwestern United States. *Ecological Applications, 15,* 154–174.

Geils, B. W., Tovar, J. C., & Moody, B. (2002). *Mistletoes of North American conifers* (General Technical Report RMRS-GTR-98). Ogden: U.S. Forest Service, Rocky Mountain Research Station.

Goheen, D. J., & Hansen, E. M. (1993). Effects of pathogens and bark beetles on forests. In T. D. Schowalter & G. M. Filip (Eds.), *Beetle-pathogen interactions in conifer forests* (pp. 75–191). San Diego: Academic.

Hagle, S. K., & Filip, G. M. (2010). *Schweinitzii root and butt rot of western conifers* (Forest Insect and Disease Leaflet 177). Portland: U.S. Forest Service, Pacific Northwest Region, State and Private Forestry, Forest Health Protection.

Hagle, S. K., Schwandt, J. W., Johnson, T. L., et al. (2000). *Successional functions of forests and pathogens, volume 2: Results* (FHP Report 00–11). Missoula: U.S Forest Service Northern Region, Forest Health Protection.

Hansen, E. M. (2014). Forest development and carbon dynamics after mountain pine beetle outbreaks. *Forest Science, 60,* 476–488.

Hansen, E. M., Bentz, B. J., & Turner, D. L. (2001). Physiological basis for flexible voltinism in the spruce beetle (Coleoptera: Scolytidae). *The Canadian Entomologist, 133,* 805–817.

Hansen, E. M., Bentz, B. J., Powell, J. A., et al. (2011). Prepupal diapause and instar IV development rates of spruce beetle, *Dendroctonus ruifpennis* (Coleoptera: Curculionidae, Scolytinae). *Journal of Insect Physiology, 57,* 1347–1357.

Hart, S. J., Veblen, T. T., Eisenhart, K. S., et al. (2013). Drought induces spruce beetle (*Dendroctonus rufipennis*) outbreaks across northwestern Colorado. *Ecology, 95,* 930–939.

Harte, J., & Shaw, R. (1995). Shifting dominance within a montane vegetation community: Results of a climate-warming experiment. *Science, 267,* 876–880.

Hawksworth, F. G., & Wiens, D. (1996). *Dwarf mistletoes: Biology, pathology, and systematics.* Washington, DC: U.S. Forest Service.

Hebertson, E. G., & Jenkins, M. J. (2007). The influence of fallen tree timing on spruce beetle brood production. *Western North American Naturalist, 67,* 452–460.

Helfer, S. (2014). Rust fungi and global change. *New Phytologist, 201,* 770–780.

Hejl, S. J., Hutto, R. L., Preston, C. R., & Finch, D. M. (1995). Effects of silvicultural treatments in the Rocky Mountains. In T. E. Martin, & D. M. Finch (Eds.), *Ecology and management of Neotropical migratory birds, a synthesis and review of critical issues* (pp. 220–224). New York: Oxford University Press.

Heyerdahl, E. K., McKenzie, D., Daniels, L. D., et al. (2008a). Climate drivers of regionally synchronous fires in the inland Northwest (1651–1900). *International Journal of Wildland Fire, 17,* 40–49.

Heyerdahl, E. K., Morgan, P., & Riser, J. P. (2008b). Multi-season climate synchronized historical fires in dry forests (1650–1900), northern Rockies, USA. *Ecology, 89,* 705–716.

Hicke, J. A., Johnson, M. C., Hayes, J. L., & Preisler, H. K. (2012). Effects of bark beetle-caused tree mortality on wildfire. *Forest Ecology and Management, 271,* 81–90.

Hicke, J. A., Meddens, A. J. H., Allen, C. D., & Kolden, C. A. (2013). Carbon stocks of trees killed by bark beetles and wildfire in the western United States. *Environmental Research Letters, 8,* 1–8.

Hoff, R., & Hagle, S. (1990). Diseases of whitebark pine with special emphasis on white pine blister rust. In: W.C. Schmidt, & K.J. McDonald (Compilers), *Symposium on whitebark pine ecosystems: Ecology and management of a high-mountain resource* (General Technical Report INT-GTR-270, pp. 179–190). Ogden: U.S. Forest Service, Intermountain Research Station.

Hoff, R., Bingham, R. T., & McDonald, G. I. (1980). Relative blister rust resistance of white pines. *European Journal of Forest Pathology, 10*, 307–316.

Hoffman, C. M., Morgan, P., Mell, W., et al. (2013). Surface fire intensity influences simulated crown fire behavior in lodgepole pine forests with recent mountain pine beetle-caused tree mortality. *Forest Science, 59*, 390–399.

Holsinger, L., Keane, R. E., Isaak, D. J., et al. (2014). Relative effects of climate change and wildfires on stream temperatures: A simulation modeling approach in a Rocky Mountain watershed. *Climatic Change, 124*, 191–206.

Jenkins, S. E., Sieg, C. H., Anderson, D. E., et al. (2011). Late Holocene geomorphic record of fire in ponderosa pine and mixed-conifer forests, Kendrick Mountain, northern Arizona, USA. *International Journal of Wildland Fire, 20*, 125–141.

Keane, R. E. (2013). Disturbance regimes and the historical range of variation in terrestrial ecosystems. In A. L. Simon (Ed.), *Encyclopedia of biodiversity* (2nd ed., pp. 568–581). Waltham: Academic.

Keane, R. E., Agee, J. K., Fulé, P., et al. (2008). Ecological effects of large fires on US landscapes: Benefit or catastrophe? *International Journal of Wildland Fire, 17*, 696–712.

Keane, R. E., Tomback, D. F., Aubry, C. A., et al. (2012). *A range-wide restoration strategy for whitebark pine forests* (General Technical Report RMRS-GTR-279). Fort Collins: U.S. Forest Service, Rocky Mountain Research Station.

Keane, R. E., Loehman, R., Clark, J., et al. (2015a). Exploring interactions among multiple disturbance agents in forest landscapes: Simulating effects of fire, beetles, and disease under climate change. In A. H. Perera, B. R. Surtevant, & L. J. Buse (Eds.), *Simulation modeling of forest landscape disturbances* (pp. 201–231). New York: Springer.

Kegley, S. J., Livingston, R. L., & Gibson, K. E. (1997). *Pine engraver, Ips pini (say), in the western United States, Forest insect and disease leaflet 122*. Washington, DC: U.S. Forest Service.

Kendall, K., & Keane, R. E. (2001). The decline of whitebark pine. In D. Tomback, S. F. Arno, & R. E. Keane (Eds.), *Whitebark pine communities: Ecology and restoration* (pp. 123–145). Washington, DC: Island Press.

Kile, G. A., McDonald, G. I., & Byler, J. W. (1991). Ecology and disease in natural forests. In C. G. Shaw III & G. A. Kile (Eds.), *Armillaria root disease* (pp. 102–121). Washington, DC: U.S. Forest Service.

Kipfmueller, K. F., & Baker, W. L. (1998). Fires and dwarf mistletoe in a Rocky Mountain lodgepole pine ecosystem. *Forest Ecology and Management, 108*, 77–84.

Kitzberger, T., Brown, P., Heyerdahl, E., et al. (2007). Contingent Pacific–Atlantic Ocean influence on multicentury wildfire synchrony over western North America. *Proceedings of the National Academy of Sciences, USA, 104*, 543–548.

Kliejunas, J. T. (2011). *A risk assessment of climate change and the impact of forest diseases on forest ecosystems in the western United States and Canada* (General Technical Report PSW-GTR-236). Albany: U.S. Forest Service, Pacific Southwest Research Station.

Kliejunas, J. T., Geils, B. W., Glaeser, J. M., et al. (2009). *Review of literature on climate change and forest diseases of western North America* (General Technical Report PSW-GTR-225). Albany: U.S. Forest Service, Pacific Southwest Research Station.

Klopfenstein, N. B., Kim, M.-S., Hanna, J. W., et al. (2009). *Approaches to predicting potential impacts of climate change on forest disease: An example with Armillaria root disease* (Research Paper RMRS-RP-76). Fort Collins: U.S. Forest Service, Rocky Mountain Research Station.

Koteen, L. (1999). Climate change, whitebark pine, and grizzly bears in the greater Yellowstone ecosystem. In S. H. Schneider & T. L. Rook (Eds.), *Wildlife responses to climate change* (pp. 343–364). Washington, DC: Island Press.

Littell, J. S., McKenzie, D., Peterson, D. L., & Westerling, A. L. (2009). Climate and wildfire area burned in western US ecoprovinces, 1916-2003. *Ecological Applications, 19*, 1003–1021.

Littell, J. S., Oneil, E. E., McKenzie, D., et al. (2010). Forest ecosystems, disturbance, and climatic change in Washington state, USA. *Climatic Change, 102*, 129–158.

Littke, W. R., & Gara, R. I. (1986). Decay of fire-damaged lodgepole pine in south-central Oregon. *Forest Ecology and Management, 17*, 279–287.

Livingston, R. L. (1979). *The pine engraver, Ips pini (say), in Idaho: Life history, habits and management recommendations*, Report 79–3. Boise: Idaho Department of Lands, Forest Insect and Disease Control.

Lockman, B., & Hartless, C. (2008). *Thinning and pruning ponderosa pine for the suppression and prevention of Elytroderma needle disease on the Bitterroot National Forest* (Region 1 Publication 08–03). Missoula: U.S. Forest Service, Northern Region, State and Private Forestry, Forest Health Protection.

Loehman, R. A., Clark, J. A., & Keane, R. E. (2011). Modeling effects of climate change and fire management on western white pine (*Pinus monticola*) in the northern Rocky Mountains, USA. *Forests, 2*, 832–860.

Logan, J. A., Macfarlane, W. W., & Willcox, L. (2008). Effective monitoring as a basis for adaptive management: A case history of mountain pine beetle in greater Yellowstone ecosystem whitebark pine. *Forest-Biogeosciences and Forestry, 2*, 19–22.

Lubchenco, J., & Karl, T. R. (2012). Predicting and managing extreme weather events. *Physics Today, 65*, 31.

Macfarlane, W. W., Logan, J. A., & Kern, W. R. (2013). An innovative aerial assessment of greater Yellowstone ecosystem mountain pine beetle-caused whitebark pine mortality. *Ecological Applications, 23*, 421–437.

Mangla, S., Sheley, R. L., James, J. J., & Radosevich, S. R. (2011). Role of competition in restoring resource poor arid systems dominated by invasive grasses. *Journal of Arid Environments, 75*, 487–493.

Marlon, J. R., Bartlein, P., Carcaillet, C., et al. (2008). Climate and human influences on global biomass burning over the past two millennia. *Nature Geoscience, 1*, 697–702.

Marlon, J. R., Bartlein, P. J., Walsh, M. K., et al. (2009). Wildfire responses to abrupt climate change in North America. *Proceedings of the National Academy of Sciences, USA, 106*, 2519–2524.

McKenzie, D., Gedalof, Z., Peterson, D., & Mote, P. (2004). Climatic change, wildfire, and conservation. *Conservation Biology, 18*, 890–902.

Mitchell, R. G., & Preisler, H. K. (1998). Fall rate of lodgepole pine killed by the mountain pine beetle in central Oregon. *Western Journal of Applied Forestry, 13*, 23–26.

Morgan, P., Heyerdahl, E. K., & Gibson, C. E. (2008). Multi-season climate synchronized forest fires throughout the 20th century, northern Rockies, USA. *Ecology, 89*, 717–728.

Ortega, Y. K., & Pearson, D. E. (2005). Strong versus weak invaders of natural plant communities: Assessing invasibility and impact. *Ecological Applications, 15*, 651–661.

Ortega, Y. K., Pearson, D. E., Waller, L. P., et al. (2012). Population-level compensation impedes biological control of an invasive forb and indirect release of a native grass. *Ecology, 93*, 783–792.

Overpeck, J. T., Rind, D., & Goldberg, R. (1990). Climate-induced changes in forest disturbance and vegetation. *Nature, 343*, 51–53.

Pearson, D. E., & Fletcher Jr., R. J. (2008). Mitigating exotic impacts: Restoring native deer mouse populations elevated by an exotic food subsidy. *Ecological Applications, 18*, 321–334.

Pelz, K. A., & Smith, F. W. (2012). Thirty year change in lodgepole and lodgepole/mixed conifer forest structure following 1980s mountain pine beetle outbreak in western Colorado, USA. *Forest Ecology and Management, 280*, 93–102.

Peterson, D. L., Johnson, M. C., Agee, J. K., et al. (2005). *Forest structure and fire hazard in dry forests of the western United States* (General Technical Report PNW-GTR-628). Portland: U.S. Forest Service, Pacific Northwest Reasearch Station.

Powell, J. A., & Bentz, B. J. (2009). Connecting phenological predictions with population growth rates for mountain pine beetle, an outbreak insect. *Landscape Ecology, 24*, 657–672.

Raffa, K., Aukema, B., Bentz, B., et al. (2008). Cross-scale drivers of natural disturbance prone to drivers of natural disturbances prone to anthropogenic amplification: The dynamics of bark beetle eruptions. *Bioscience, 58*, 501–517.

Régnière, J., & Bentz, B. (2007). Modeling cold tolerance in the mountain pine beetle, *Dendroctonus ponderosae*. *Journal of Insect Physiology, 53*, 559–572.

Régnière, J., Bentz, B. J., Powell, J. A., & St-Amant, R. (2015). Individual based modeling: Mountain pine beetle seasonal biology in response to climate. In A. H. Perera, B. Sturtevant, & L. J. Buse (Eds.), *Simulation modeling of forest landscape disturbances* (pp. 135–164). Basel: Springer.

Regonda, S. K., Rajagopalan, B., Clark, M., & Pitlick, J. (2005). Seasonal cycle shifts in hydroclimatology over the western United States. *Climate, 18*, 372–384.

Reinhardt, E. D., Holsinger, L., & Keane, R. E. (2010). Effects of biomass removal treatments on stand-level fire characteristics in major forest types of the northern Rocky Mountains. *Western Journal of Applied Forestry, 25*, 34–41.

Riley, K., & Loehman, R. (2016). Mid-21st century climate changes increase predicted fire occurrence and fire season length, northern Rocky Mountains, US. *Ecosphere, 7*, e01453.

Rocca, M. E., Brown, P. M., MacDonald, L. H., & Carrico, C. M. (2014). Climate change impacts on fire regimes and key ecosystem services in Rocky Mountain forests. *Forest Ecology and Management, 327*, 290–305.

Safranyik, L., & Linton, D. A. (1991). Unseasonably low fall and winter temperatures affecting mountain pine beetle and pine engraver beetle populations and damage in the British Columbia Chilcotin region. *Journal of the Entomological Society of British Columbia, 88*, 17–21.

Schoennagel, T. L., Veblen, T. T., & Romme, W. H. (2004). The interaction of fire, fuels, and climate across Rocky Mountain landscapes. *Bioscience, 54*, 651–672.

Seidl, R., Fernandes, P. M., Fonseca, T. F., et al. (2011). Modelling natural disturbances in forest ecosystems: A review. *Ecological Modelling, 222*, 903–924.

Shore, T., Safranyik, L., Riel, W., et al. (1999). Evaluation of factors affecting tree and stand susceptibility to the Douglas-fir beetle (Coleoptera: Scolytidae). *The Canadian Entomologist, 131*, 831–839.

Smith, R. S., Jr. (1989). History of *Heterobasidion annosum* in western United States. In: W. J. Otrosina, & R. F. Scharpf (Technical coordinators), *Proceedings of the symposium on research and management of annosus root disease* (Heterobasidion annosum) in western North America (General Technical Report GTR-PSW-116; pp. 10–16). Berkeley: U.S. Forest Service, Pacific Southwest Forest and Range Experiment Station.

Spracklen, D., Mickley, L., Logan, J., et al. (2009). Impacts of climate change from 2000 to 2050 on wildfire activity and carbonaceous aerosol concentrations in the western United States. *Journal of Geophysical Research, 114*, D20301.

Stephens, S. L., & Finney, M. (2002). Prescribed fire mortality of Sierra Nevada mixed conifer tree species: Effects of crown damage and forest floor combustion. *Forest Ecology and Management, 162*, 261–271.

Stone, J. K., Coop, L. B., & Manter, D. K. (2008). Predicting effects of climate change on Swiss needle cast disease severity in Pacific Northwest forests. *Canadian Journal of Plant Pathology, 30*, 169–176.

Sturrock, R. N., Frankel, S. J., Brown, A. V., et al. (2011). Climate change and forest diseases. *Plant Pathology, 60*, 133–149.

Swetnam, T. W., Allen, C. D., & Betancourt, J. L. (1999). Applied historical ecology: Using the past to manage for the future. *Ecological Applications, 9*, 1189–1206.

Taylor, A. H., & Skinner, C. N. (2003). Spatial patterns and controls on historical fire regimes and forest structure in the Klamath Mountains. *Ecological Applications, 13*, 704–719.

Thuiller, W., Albert, C., Araújo, M. B., et al. (2008). Predicting global change impacts on plant species' distributions: Future challenges. *Perspectives in Plant Ecology, Evolution and Systematics, 9*, 137–152.

Tilman, D., & El Haddi, A. (1992). Drought and biodiversity in grasslands. *Oecologia, 89*, 257–264.

Tomback, D. F., & Achuff, P. (2011). Blister rust and western forest biodiversity: Ecology, values and outlook for white pines. *Forest Pathology, 40*, 186–225.

U.S. Forest Service (USFS). (2007). *Forest insect and disease conditions in the United States - 2006*. Washington, DC: U.S. Forest Service, Forest Health Protection.

Walther, G. R., Roques, A., Hulme, P. E., et al. (2009). Alien species in a warmer world: Risks and opportunities. *Trends in Ecology and Evolution, 24*, 686–693.

Weed, A. S., Bentz, B. J., Ayres, M. P., & Holmes, T. P. (2015). Geographically variable response of *Dendroctonus ponderosae* to winter warming in the western United States. *Landscape Ecology, 30*, 1075–1093.

Westerling, A. L., & Swetnam, T. W. (2003). Interannual to decadal drought and wildfire in the western United States. *Eos, Transactions American Geophysical Union, 84*, 545.

Westerling, A. L., Hidalgo, H. G., Cayan, D. R., & Swetnam, T. W. (2006). Warming and earlier spring increase in western US forest wildfire activity. *Science, 313*, 940–943.

Westerling, A. L., Turner, M. G., Smithwick, E. A. H., et al. (2011). Continued warming could transform greater Yellowstone fire regimes by mid-21st century. *Proceedings of the National Academy of Sciences, USA, 108*, 13165–13170.

White, P. S., & Pickett, S. T. A. (1985). Natural disturbance and patch dynamics: An introduction. In S. T. A. Pickett & P. S. White (Eds.), *The ecology of natural disturbance and patch dynamics* (pp. 3–13). Orlando: Academic.

Williams, A. P., Allen, C. D., Macalady, A. D., et al. (2013). Temperature as a potent driver of regional forest drought stress and tree mortality. *Nature Climate Change, 3*, 292–297.

Zambino, P. J. (2010). Biology and pathology of ribes and their implications for management of white pine blister rust. *Forest Pathology, 40*, 264–291.

Zimmerman, G. T., & Laven, R. D. (1987). Effects of forest fuel smoke on dwarf mistletoe seed germination. *Great Basin Naturalist, 47*, 652–659.

Zouhar, K., Smith, J.K., Sutherland, S., & Brooks, M.L. (2008). *Wildland fire in ecosystems: Fire and nonnative invasive plants* (General Technical Report RMRS-GTR-42). Ogden: U.S. Forest Service. Rocky Mountain Research Station.

Chapter 8
Effects of Climate Change on Wildlife in the Northern Rockies

Kevin S. McKelvey and Polly C. Buotte

Abstract Few data exist on the direct effects of climatic variability and change on animal species. Therefore, projected climate change effects must be inferred from what is known about habitat characteristics and the autecology of each species. Habitat for mammals, including predators (Canada lynx, fisher, wolverine) and prey (snowshoe hare) that depend on high-elevation, snowy environments, is expected to deteriorate relatively soon if snowpack continues to decrease. Species that are highly dependent on a narrow range of habitat (pygmy rabbit, Brewer's sparrow, greater sage-grouse) will be especially vulnerable if that habitat decreases from increased disturbance (e.g., sagebrush mortality from wildfire). Species that are mobile or respond well to increased disturbance and habitat patchiness (deer, elk) will probably be resilient to a warmer climate in most locations. Some amphibian species (Columbia spotted frog, western toad) may be affected by pathogens (e.g., amphibian chytrid fungus) that are favored by a warmer climate.

Adaptation strategies for wildlife focused on maintaining adequate habitat and healthy wildlife populations, and increasing knowledge of species' needs and climate sensitivities. Connectivity is an important conservation strategy for most species in the Northern Rockies. Maintaining healthy American beaver populations will provide riparian habitat structure and foraging opportunities for multiple species. Quaking aspen habitat, which is also important for several species, can be enhanced by allowing wildfire to burn, protecting aspen from grazing, and reducing conifer encroachment. Restoration of more open stands of ponderosa pine and mixed conifer forest through reduction of stand densities will benefit species such as flammulated owl. Excluding fire and reducing nonnative species will maintain sagebrush habitats that are required by several bird and mammal species.

K.S. McKelvey (✉)
U.S. Forest Service, Rocky Mountain Research Station, Missoula, MT, USA
e-mail: kmckelvey@fs.fed.us

P.C. Buotte
College of Forestry, Oregon State University, Corvallis, OR, USA
e-mail: pcbuotte@gmail.com

Keywords Wildlife • Climate change • Adaptation • Northern Rockies • Fisher • Pygmy rabbit • Brewer's sparrow • Deer • Elk • Columbia spotted frog • Western toad • Connectivity • American beaver • Lynx • Wolverine • Sage-grouse • Riparian • Wetland • Pelage change

8.1 Climate-Wildlife Interactions

Temperature and moisture affect animal physiological response at short time scales via the thin boundary layer immediately above their tissues (Fig. 8.1). If you (a mammal) are wearing dark clothing on a cold, sunny day, sun energy interacts with the dark clothing, creating a warm boundary layer. Conditions beyond that thin boundary layer are physiologically irrelevant. In the shade, the warm boundary layer is replaced with one at the ambient temperature of the air, making you cold. This example demonstrates a number of basic factors that need to be considered when assessing the effects of climate change on animals. Climate is, by definition, the long-term composite of weather, which in turn is the composite of these nearly instantaneous effects in an organism's environment. Climate changes the frequency of weather events, which in turn changes the frequency of fast shifts in boundary layer conditions. But organisms do not directly respond to climate and can seek to optimize changes in climatic conditions through metabolic and behavioral plasticity.

Terrestrial animals can manipulate their environment by standing in the sun or shade, moving uphill or downhill, changing aspect, or seeking cooler/warmer conditions by digging into a burrow or the snow. Endothermic animals change their boundary layer by modifying hair or feathers, seasonally and at much shorter time scales, while minimizing energy expenditure. Endotherms can further regulate their body temperatures by expending energy; changes in climate may be expressed as increased

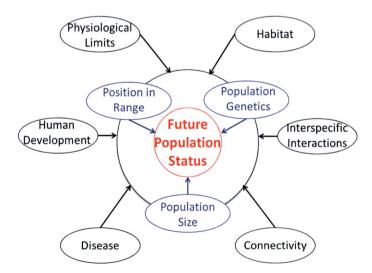

Fig. 8.1 Conceptual diagram of the effects of climate on wildlife populations in the Northern Rockies. Climate pathways (*black*) interact with population characteristics (*blue*) to affect population status (*red*)

metabolic demands. Ectotherms have no ability to regulate their body temperatures through metabolic processes; they are primarily limited to seeking appropriate temperatures by, for example, inhabiting burrows or caves (Box 8.1). It is more difficult for aquatic organisms to avoid adverse temperatures because water conducts heat efficiently, and aquatic ectotherms are particularly vulnerable. Aquatic ectotherms cannot avoid overheating when water temperatures increase, so it is more straightforward to evaluate climate change effects on fish with known warm-water limits (Chap. 4).

Because of their inherent plasticity when faced with changing temperature, terrestrial endotherms are more likely to experience effects associated with changes in precipitation amounts and types, because water produces physical features that serve as habitat. In the Northern Rockies, where winters are cold, snow provides physical habitat for which some animals have specific adaptations, such as a seasonal color change in pelage: white to match snow, brown to match a snowfree background (see discussion of snowshoe hare [*Lepus americanus*] below) (Fig. 8.2).

> **Box 8.1 How Animals Respond to Temperature: Endotherms vs. Ectotherms**
>
> **Endotherms** (warm-blooded animals) maintain a constant body temperature, and cold or excessive heat requires them to burn more calories to maintain a core temperature, allowing them to function in a wide variety of environmental conditions. Endotherm physiology responds directly to temperature change, and effectiveness and sustainability of physiological response are determined by the quantity and quality of available food.
>
> **Ectotherms** (cold-blooded animals), which include fish, reptiles, and amphibians, react not by feeling cold and metabolizing energy to maintain core temperature, but by having their metabolism slow until their activity is reduced, in some cases becoming torpid. These basic metabolic differences in vertebrates must be considered in how climate change will interact with animal life history, spatial distribution, habitat quality, and food sources.

Fig. 8.2 Canada lynx diet is dominated by snowshoe hare, which undergoes seasonal pelage change from brown to white. The effectiveness of this strategy depends on synchrony with snow cover (*Left* photo by Milo Burcham, *right* photo by L. Scott Mills)

Table 8.1 Species included in the Northern Rockies Adaptation Partnership vulnerability assessment, summarized by subregion

Habitat/species	Western Rockies	Central Rockies	Eastern Rockies	Greater Yellowstone Area	Grassland
Dry forest					
Flammulated owl		X		X	
Pygmy nuthatch		X	X	X	
Riparian/wetland					
American beaver		X	X	X	
Moose				X	
Northern bog lemming				X	
Townsend's big-eared bat		X	X	X	
Harlequin duck		X		X	
Columbia spotted frog		X		X	
Western toad		X		X	
Quaking aspen					
Ruffed grouse				X	
Sagebrush grasslands					
Pronghorn				X	
Pygmy rabbit			X		
Brewer's sparrow				X	
Greater sage-grouse				X	X
Mountain grasslands					
Mountain quail	X				
Mesic old-growth forest					
Fisher		X		X	
Snow-dependent species					
American pika				X	
Canada lynx		X		X	
Wolverine		X		X	
Ungulates: elk, mule deer, white-tailed deer	X	X	X	X	X

Oversized feet, long legs, and light bone structure also provide benefits in snow-covered landscapes.

Deep snow provides a warm, stable environment at the interface between snow and soil, and soil temperatures can often remain above freezing throughout winter (Edwards et al. 2007). For animals that depend on a stable subnivean environment, or who have specific phenological adaptations to snow, reduced snowpack (Chap. 3) would represent a loss of critical habitat. Similarly, water bodies provide physical habitats with features that provide predator avoidance, temperature control, and sources of food. In addition, open or flowing water can provide important microclimates. For example, American pikas (*Ochotona princeps*) can be found in otherwise hot environments if water flows beneath the talus, producing cool microsites (Millar and Westfall 2010). Seeps, springs, bogs, and persistent streams can integrate longer climatic periods, so altering these features can make species that depend on them vulnerable to climate change (Table 8.1).

Climate effects on terrestrial endotherms will often be a function of changes in plant assemblages that comprise wildlife habitat (Fig. 8.1). For predators, these effects may be either direct (changes in number and location of vegetation boundaries used by predators) or indirect (changes in prey densities or prey availability to predators). If effects are associated with changes in habitat, projecting climate effects on a specific animal species will be difficult, requiring knowledge about the functional roles of ecological attributes in an animal's life history, and consequences associated with different life history strategies. Current behaviors can be studied, but may not be informative about climate change effects, and responses may be novel or difficult to anticipate.

Trophic effects include presence and abundance of disease and parasitic organisms. For example, for greater sage-grouse (*Centrocercus urophasianus*), the potential spread of West Nile virus (*Flavivirus* spp.) associated with climate change may increase stress in grouse populations (Schrag et al. 2011). For many organisms, current ranges are often strongly limited by human activities. For example, the range of greater sage-grouse is limited by conversion of native sagebrush (*Artemisia* spp.) habitat to agricultural uses (Miller and Eddleman 2001). In addition, climate change will alter the nature and location of human activities that affect wildlife. In the western United States, changes in water availability and uses will have major effects on human settlement patterns (Barnett et al. 2005), which will in turn affect habitat.

8.2 Communities and Habitat

Our understanding of wildlife ecology, particularly at broad spatial scales, is generally limited to relationships between occurrence patterns rather than direct studies of limiting factors. Although patterns of occurrence may be clear, consistent, and correlated with climate, causal mechanisms may be difficult to infer. For example, many passerine birds nest only in specific habitats, such as Brewer's sparrow (*Spizella breweri*) which is obligate to sagebrush. The pattern is clear and invariant, but the mechanistic links with sagebrush are unknown. Species such as ruffed grouse (*Bonasa umbellus*) have northern distributions, but limiting factors for its southern distribution are poorly understood (Lowe et al. 2010).

Based on observed patterns of distribution, enough information exists to identify and manage current habitat, but it cannot be assumed that measured correlations will persist in an altered climate. Assume that an animal's occurrence is strongly correlated with mature Douglas-fir (*Pseudotsuga menziesii*) forests. These forests contain other tree and understory species, animal communities, and successional trajectories, but in a future landscape, Douglas-fir may be associated with different plant and animal communities. The correlational nature of most habitat data makes it difficult to know which components are critical to habitat quality of a certain species, much less habitat quality of novel species assemblages.

The effects of climate on future habitats are expected to be strongly influenced by altered disturbance regimes. Changing disturbance dynamics (Chap. 7) modify characteristics of landscape patterns across a broad range of spatial scales. If climate change causes shifts in plant and animal distribution and abundance, a temporal mismatch may occur between decreased current habitat and increased new habitat, and will be exacerbated by periodic disturbance. Wildfire can destroy current habitat in a day, but generation of new habitat may require centuries. Fisher (*Pekania pennanti*) provides an example of uncertainties associated with projecting the effects of disturbance. This species is currently limited to mature forests in the Inland Maritime climatic zone of Idaho and Montana, and climate projections indicate that this zone will move to the east. Although climatic conditions to the east may be similar to those in areas occupied by fisher, the habitat associated with mature forest requires a century or more to grow in the projected climate zone, and a projected increase in disturbances may prevent that from occurring (Chap. 7). Therefore, it is unclear whether the projected climate will actually provide fisher habitat.

Because trajectories of species and their habitats under climate change are uncertain, we consider vulnerability assessments for animals as hypotheses to be tested. We assume that proactive management strategies will be used in the future to maintain valued species and landscape attributes, including creating resilience to disturbance. A monitoring program designed to test specific hypotheses associated with specific organisms (Nichols and Williams 2006) can improve our understanding of relationships between climate change and landscapes, providing data that inform science-based management.

8.3 Species Sensitivity to Climate Change

A few animal species have received significant attention, generating peer-reviewed articles that analyze the effects of climate change, although this is relatively uncommon. Foden et al. (2013) identified three dimensions associated with climate change vulnerability—sensitivity, exposure, and adaptive capacity—and applied a framework based on assessing these attributes to nearly 17,000 species. Other expert systems have been developed to evaluate the relative degree of climate sensitivity and vulnerability for various species. These tools do not seek to understand specific responses of animals to climate, but rather to identify species that are likely to be vulnerable based on current habitat associations, life history traits, and distributions (Foden et al. 2013). Bagne et al. (2011) formalized this process in the System for Assessing Vulnerability of Species (SAVS), which considers a large number of traits associated with habitat (7 traits), physiology (6 traits), phenology (4 traits), and biotic interactions (5 traits).

Formalizing traits that can lead to vulnerability provides a framework for collecting biological data associated with a species and for considering the effects of climate change. However, existing expert systems cannot be used to infer the relative importance of disparate sources of vulnerability such as habitat and phenology or if estimated vulnerability scores have quantitative meaning (Bagne et al. 2011; Case et al. 2015). For most species, accurately identifying vulnerability (Foden et al. 2013) would be challenging given current biological understanding. Because published data on climate-species relationships are so sparse, we focus on evaluation of each trait as it relates to the biology of specific animal species.

Below are assessments for animal species identified as high priority by U.S. Forest Service Northern Region and national forest resource specialists. Species were not necessarily chosen based on their perceived level of vulnerability. In many cases, species are associated with specific habitats that were considered vulnerable; for example, some species are associated with sagebrush communities, others with snow depth and cover, and others with dry forests that have large trees. Inferences are based on interpretation of the pertinent literature, including empirical data, modeling, and autecology. Level of detail differs greatly among species, proportional to the amount of information available.

8.3.1 American Beaver (Castor canadensis)

American beavers spend most of the winter in lodges or swimming to retrieve food, so climate may be more influential during spring through autumn than during winter (Jarema et al. 2009). However, body weights of juvenile European beavers were lighter when winters were colder. The cost of thermodynamic regulation may be greater for juveniles because they have higher surface area-to-volume ratios than adults (Campbell et al. 2013). In Québec (Canada), beaver density was highest in areas with highest maximum spring and summer temperatures (Jarema et al. 2009). Conversely, European beavers in Norway achieved heavier body weights when spring temperatures were lower, and the rate of vegetation green-up was slower (Campbell et al. 2013). This apparent contradiction may have been caused by the timing and measurement of climate and response variables. Although beavers require ponds, survival and body weight in European beavers has been linked to lower April–September precipitation (Campbell et al. 2012, 2013).

Climate can indirectly influence beavers through effects on vegetation. Climate change and climate-driven changes in streamflow may reduce early-seral tree

Fig. 8.3 Restoration of American beaver populations helps maintain cool water in mountain landscapes. Beavers create structures that help ameliorate the effects of climate change on habitat for cold-water fish species and other aquatic organisms (Photo by E. Himmel, National Park Service)

species in riparian habitats (Perry et al. 2012), thus reducing food and building materials. Beavers can be used to buffer riparian systems from drought, because beaver ponds increase the amount of open water, assisting the conservation of other animals, such as amphibians (Fig. 8.3) (Chaps. 3, 4).

8.3.2 American Pika (Ochotona princeps)

American pika is a small lagomorph that inhabits rocky alpine areas in western North America (Smith and Weston 1990). Relatively little study of pikas had occurred in the Northern Rockies until recently, with the exception of research on occupancy and abundance in the Bighorn Mountains and Wind River Range (Wyoming) (Yandow 2013). Pikas depend on moist, cool summer conditions and winter snow (Beever et al. 2011), and on low water-balance stress and green vegetation (Beever et al. 2013). Acute temperature stresses (hot and cold) and vegetation productivity appear to affect pika declines in the Great Basin (Beever et al. 2010, 2011, 2013), reinforcing surveys in Colorado of 4 pika extirpations (of 69 total sites) that occurred at the driest sites (Erb et al. 2011).

Winter snowpack insulates pikas during cold periods and provides water during summer. Surveys in the Sierra Nevada found that pika extirpations were associated with sites with higher maximum temperatures and lower annual precipitation (Millar and Westfall 2010). Individual mountain ranges act as discrete areas without pika migration between adjacent ranges across valley bottoms (Castillo et al. 2014). Connectivity of pika populations appears to be context dependent, with lower connectivity between sites that occur in hotter, drier landscapes (Henry et al. 2012; Castillo et al. 2014). Thus, recolonization may occur at distances less than 0.8 km and in areas where between-population dispersal occurs within cool, moist landscapes.

Studies in the Sierra Nevada (Millar and Westfall 2010) and southern Rocky Mountains (Erb et al. 2011), at sites in which pikas were common and not generally subject to extirpation across most of the landscape, indicated that physiological limits for this species had not been reached. This will probably be the case for most pika populations in the Northern Rockies in the near term. Existence of pikas at Lava Beds National Monument, Craters of the Moon National Monument, and the Columbia River Gorge—all of which have warm, dry climates—illustrates the importance of microclimate for suitable habitat. Because pikas are sensitive to high temperatures, their populations will probably respond to climate change in the Northern Rockies. However, site-specific factors contribute to highly variable microclimates, so response to climate change will probably be minimal and vary over space and time.

8.3.3 Canada Lynx (Lynx canadensis)

Canada lynx is a mid-sized cat with adaptations that allow it to travel across soft snow, including oversized feet. Canada lynx prey nearly exclusively on snowshoe hare (*Lepus americanus*) (Fig. 8.2), which constitutes 33–100% of its diet (Mowat et al. 2000; Squires and Ruggiero 2007). Snowshoe hares also exhibit seasonal pelage change from brown to white.

Lynx are found exclusively in North America, extending across interior Canada and Alaska and northward into tundra vegetation and southward into high mountain regions in the conterminous United States (McKelvey et al. 2000). In the Northern Rockies, lynx exist in only the Clearwater River watershed, Bob Marshall Wilderness, and northwestern corner of Montana. Maintaining population connectivity is central to lynx conservation. However, maintaining connectivity may become increasingly difficult as southern populations of boreal species become more isolated with climate change (van Oort et al. 2011), especially where disturbance processes are expected to increase.

In the Northern Rockies, 80% of dens are in mature forest and 13% in mid-seral regenerating stands (Squires et al. 2008). For winter foraging, lynx forage in mature, multilayer spruce-fir forests composed of large diameter trees with high horizontal cover, abundant snowshoe hares, and deep snow (Squires et al. 2010). Lynx select home ranges with vegetative conditions consistent with those identified for foraging and denning (Squires et al. 2013). The range of snowshoe hare is more extensive than that of lynx, extending into the mid Sierra Nevada and areas such as the Olympic Peninsula (McKelvey et al. 2000). The more extensive hare distribution, which includes areas with limited snow (e.g., the Pacific coast), is likely associated with greater genetic differentiation in snowshoe hares compared to lynx.

Variation in timing of pelage change in snowshoe hares is low in any specific location, and timing appears to be genetically controlled and linked to photoperiod (Zimova et al. 2014). Timing of pelage change is critical, because mismatches—a white hare on a dark background and vice versa—make hares susceptible to predation (Hodges 2000) (Fig. 8.2), and the ability of hares to shift timing of pelage change to match snow patterns is limited. Unless a significant change occurs in the population genetics of hares, they will be the wrong color for approximately 2 months per year in the Northern Rockies (Mills et al. 2013). Both lynx and hares require specific amounts and duration of winter snow (McKelvey et al. 2000; Schwartz et al. 2004), and in western Montana, lynx and hares use older spruce-fir forests. If climate change and associated disturbance reduce the abundance of these forests, populations of both lynx and hares could decline significantly.

8.3.4 Fisher (Pekania pennanti)

The fisher is a mid-sized, forest-dwelling mustelid whose range covers much of the boreal forest in Canada, a broad area of the northeastern United States extending from the Lake States to Maine, and a scattered distribution in the western United States. Fishers are often associated with urban environments in the eastern United States, but this is uncommon in the West where fishers are associated with very specific habitats and populations are disjunct. Common attributes for resting sites for western fisher are steep slopes, cool microclimates, dense forest canopy cover, high volume of logs, and prevalence of large trees and snags (Aubry et al. 2013).

In the West, fishers are associated with specific habitat conditions, especially forested areas with low monthly snowfall (<13 cm) (Krohn et al. 1995), and appear to avoid both deep snowpack (Raine 1983) and dry habitats (Schwartz et al. 2013). In the Northern Rockies, fisher habitat is best in areas with high annual precipitation, low relief, and mid-range values for mean temperature in the coldest month (Olson et al. 2014). In the near term, habitat currently occupied by fisher might improve in areas that are currently occupied (primarily central Idaho) but quality will decline sharply by 2090, and new habitat may be created in northwestern Montana (Olson et al. 2014).

Given that fishers are associated with mature forests, long time lags may exist between the loss of current habitat and formation of new habitat in areas that are currently unsuitable. If large trees cannot survive the shift in climate, mature forests may become rare for many decades. In climatic zones suitable for fishers, forests may be dominated by young trees and shrubs whose suitability for fisher habitat is unknown. Projections in Olson et al. (2014) provide an optimistic view of habitat availability under climate change, especially because it is uncertain if fisher would disperse into new habitat if and when it occurs.

8.3.5 Moose (Alces alces)

Moose is an example of a well-studied animal that has a northern distribution but whose dependency on boreal environments may not be obvious. The distribution of moose is limited by food supply, climate, and habitat (Murray et al. 2006, 2012). The species is intolerant of heat, but well adapted to cold; high summer temperatures increase metabolism and heart and respiration rates, and reduce body weight (Renecker and Hudson 1986). These temperature thresholds represent physiological thresholds that, when exceeded, represent heat stress that increases the energy expenditure needed to stay cool.

Because of the behavioral plasticity discussed above, moose may be able to avoid being exposed to high mid-day summer temperatures. In Minnesota, Lenarz et al. (2009) found that temperature was highly correlated with moose survival, but winter temperature was more critical than summer heat. In northern Minnesota, moose populations were not viable, largely because of disease and parasite-related mortality (Murray et al. 2006). However, in nearby southern Ontario (Canada), moose populations had favorable growth rates (Murray et al. 2012). Warming temperatures favor white-tailed deer (*Odocoileus virginianus*) expansion into moose range, creating the potential for increased transmission of deer parasites to moose (Lankester 2010). For moose, separating direct and indirect climate effects is difficult.

8.3.6 Northern Bog Lemming (Synaptomys borealis)

The northern bog lemming inhabits wet meadows, bogs, and fens within several overstory habitat types (Foresman 2012), typically with extensive sphagnum (*Sphagnum* spp.), willow (*Salix* spp.), or sedge. Given their dependence on wet habitats, it follows that climate changes that decrease the amount of surface water will have negative impacts on lemmings. Therefore, management practices that maintain surface water may be beneficial, although documented studies of climate and management effects are lacking.

8.3.7 Pronghorn (Antilocapra americana)

The pronghorn (*Antilocapra americana*) is an ungulate native to the prairies, shrublands, and deserts of the western United States, occupying a broad range of climatic conditions from southern Canada to Mexico. Pronghorns are prone to epizootic diseases, notably bluetongue (a viral disease transmitted by midges [*Culicoides* spp.]) (Thorne et al. 1988). Bluetongue is thought to be cold-weather limited, and recent extensions of bluetongue in Europe have been attributed to climatic warming (Purse et al. 2005). Given their current range and food habits, the emergence of new disease threats caused by a warmer climate probably poses the greatest risk to pronghorns.

8.3.8 Pygmy Rabbit (Brachylagus idahoensis)

The pygmy rabbit is one of the smallest leporids in the world and is endemic to big sagebrush (*Artemisia tridentata*), which provides food and cover. In southeastern Idaho, pygmy rabbits were associated with higher woody cover and height than other areas, with fewer grasses and more forbs. Sagebrush was 51% of summer diet and 99% of winter diet (Green and Flinders 1980). Areas used by pygmy rabbits accumulate relatively high snowpack, and rabbits use the subnivean environment to access food and avoid predators. Snow for thermal protection may be important for winter survival, because of small body size, lack of metabolic torpor, and lack of food caching (Katzner and Parker 1997).

Processes that reduce the size and density of sagebrush are likely to have negative effects on pygmy rabbits, and processes that fragment sagebrush stands may decrease habitat quality. For example, Pierce et al. (2011) found that burrows, observed rabbits, and fecal pellets decreased in density with proximity (<100 m) to edges. Big sagebrush is sensitive to fire, and 100% mortality and stand replacement after burning are common (Davies et al. 2011; Chap. 6). Recruitment of mountain big sagebrush (*A. tridentata* ssp. *vaseyana*) relies on wind dispersal of seeds from adjacent seed sources and on composition of the soil seed bank. Mountain big sagebrush required 13–27 years after spring prescribed burning to return to conditions suitable for pygmy rabbit habitat (Woods et al. 2013).

Climatic variability has affected sagebrush communities and pygmy rabbits in the past (Grayson 2000), and this will probably happen again in the future. First, pygmy rabbits depend on a single species (sagebrush) and habitat condition (tall, dense stands). Second, pygmy rabbit habitat is sensitive to altered disturbance, especially wildfire. Finally, changes in winter snow depth could affect overwinter survival by altering protection provided by the subnivean environment.

8.3.9 Townsend's Big-Eared Bat (Corynorhinus townsendii)

Climate change can affect foraging, drinking water availability, and timing of hibernation in bats (Sherwin et al. 2013). Townsend's big-eared bats generally require caves for diurnal, maternal, and hibernation roosting, although they also use large tree cavities, buildings, and bridges and forage for insects along riparian and forest edge habitats (Fellers and Pierson 2002). This species does not produce concentrated urine and therefore requires daily access to water sources (Gruver and Keinath 2003).

Bioaccumulation of pesticides in fat tissue is one cause of declines in Townsend's big-eared bat populations (Clark 1988). Bats are especially sensitive to human disturbance during hibernation. In Colorado, reproductive success of *Myotis* spp. decreased during warm, dry conditions (Adams 2010), although warmer spring temperatures have led to earlier births and higher juvenile survival (Lucan et al. 2013). Higher summer precipitation may reduce reproductive success, and increased warming may reduce effectiveness of some bat echolocation calls (Luo et al. 2014).

8.3.10 Ungulates (Elk, Mule Deer, White-Tailed Deer)

Rocky Mountain elk (*Cervus canadensis*), Rocky Mountain mule deer (*Odocoileus hemionus hemionus*) and white-tailed deer (*O. virginianus*) have broad ranges in North America, indicating a high degree of habitat flexibility. Habitat use by elk in forested areas is associated with edges that contain high quality forage and nearby forest cover. In open habitats, they select areas of high vegetative diversity with intermixed patches of shrubs and grasslands (Sawyer et al. 2007). Both types of habitat are favored by disturbances with spatial heterogeneity at fine scales.

Mule deer have larger home range sizes in areas with few large patches and smaller sizes in fine-grained vegetation mosaics (Kie et al. 2002). Fine-grained disturbance mosaics are optimal for white-tailed deer, especially in areas where thermal cover is important. In the Northern Rockies, thermal cover prevents heat loss

Fig. 8.4 Ungulates generally respond positively to wildfires that create patchy habitat with improved forage, as shown in this photo of an elk browsing in a recently burned lodgepole pine forest (Photo by Jeff Henry, National Park Service)

during winter, although in warmer climates, thermal cover reduces daytime heating.

Ungulates respond positively to disturbance, with wildfire intensity affecting both species composition and patchiness in the postfire landscape (Fig. 8.4). Vegetation growth after disturbance is especially important where nonnative species are common. For example, Bergman et al. (2014) found that treatments that removed trees and controlled weeds produced better mule deer habitat than treatments that removed only trees. Climate change is expected to alter fire regimes, but for ungulates the exact nature of those changes will be critical. For example, in the Greater Yellowstone Area subregion, if climate change causes more frequent fires, then the landscape will be patchier compared to the current condition, and the distribution and abundance of forest species could change. In the long term, the effects of altered vegetation on ungulate populations are uncertain, and it is likely that there will be both positive and negative consequences.

8.3.11 Wolverine (Gulo gulo)

The wolverine is the largest mustelid, occurring throughout the Arctic, as well as subarctic areas and boreal forests of western North America and Eurasia. At the southern extent of its distribution in North America, populations occupy peninsular extensions of temperate montane forests. Wolverines den in snow, with deep snow throughout the denning period being essential (Magoun and Copeland 1998). A proxy for spring snowpack (areas where snow persisted through mid May) effectively describes den site selection, current range limits, and year-around habitat use at the southern periphery of their range (Copeland et al. 2010). Because wolverines travel within these same areas when dispersing and minimize travel through low-elevation habitat (Schwartz et al. 2009), it is possible to project current and future travel routes based on altered snowpack.

McKelvey et al. (2011) modeled future spring snowpack within the Columbia, Upper Missouri, and Colorado River basins, projecting changes in wolverine habitat and connectivity associated with future landscapes. In the Columbia and Upper Missouri River Basins, snowpack was projected to decrease 35% and 24%, respectively, for spring snow by the mid-twenty-first century, and 66% and 51%, respectively, by the end of the century. Central Idaho was projected to lose nearly all snow by the end of the century, whereas northern Montana, the southern Bitterroot Mountains, and the Greater Yellowstone Area retained significant spring snow (McKelvey et al. 2011). A connectivity model (Schwartz et al. 2009) in conjunction with ensemble climate model projections indicated that all remaining habitat blocks would likely be genetically isolated by the end of the twenty-first century (McKelvey et al. 2011). A continuing reduction in spring snow, a pattern that has been ongoing since the 1950s (Mote et al. 2005), will reduce the amount of suitable habitat for wolverines.

8.3.12 Brewer's Sparrow *(Spizella breweri)*

Brewer's sparrow is a sagebrush obligate during the nesting period, preferring tall, dense stands of sagebrush (Fig. 8.5). In many areas, Brewer's sparrows are the most abundant bird species (Norvell et al. 2014). The obligate relationship of Brewer's sparrow with sagebrush lacks causal explanations (Petersen and Best 1985). Therefore, correlative associations can be used to project climate change effects, but

Fig. 8.5 Sagebrush-obligate species such as Brewer's sparrow (shown here) may have less nesting habitat in the future if the extent of mature sagebrush habitat is reduced by wildfire (Photo by Paul Higgins, www.utahbirds.org)

we cannot speculate on the flexibility of this species to shift to other shrub species. Brewer's sparrows exhibit some flexibility in nesting locations, shifting sequential nests in response to previous predation (Chalfoun and Martin 2010).

Brewer's sparrow populations appear stable range-wide, although they have been in decline in some areas in Colorado (USGS 2013). Although Brewer's sparrows select for areas with tall, dense sagebrush, sparrow abundance was unaffected by treatments designed to modify sagebrush cover and improve habitat for greater sage-grouse (Norvell et al. 2014). In general, the effects of climate change on Brewer's sparrows will depend on changes in the distribution, abundance, composition, and structure of sagebrush communities. Increased wildfire is expected to reduce the distribution, abundance, and age of sagebrush stands in a warmer climate.

8.3.13 *Flammulated Owl* (Otus flammeolus)

The flammulated owl is a small nocturnal owl that is migratory but breeds in montane areas across much of western North America. A cavity nester associated with mature forests with large diameter trees, it is also associated with open forests composed of multiple tree species. This species is thought to be an obligate secondary cavity nester, although it has been anecdotally observed to nest in the ground (Smucker and Marks 2013). Flammulated owls feed almost exclusively on insects, primarily Lepidoptera (Linkhart et al. 1998). During the nesting period, males are single-trip, central-place foragers, so the energetics of prey selection are important, with distance traveled and energy content of prey differing by forest type.

The extensive latitudinal range of flammulated owls, lack of specific forest associations, and generalized insect diets provide few clues about sensitivity to a warmer climate. Potential effects of climate change would most likely be through disturbance processes that remove large diameter trees. Shifts to denser forest structure would be a concern, but this is unlikely because drought and wildfire are projected to increase throughout the Northern Rockies (Chap. 5). Like other long-lived owl species, flammulated owls will require a high rate of adult survival to persist in future habitats (Noon and Biles 1990).

8.3.14 *Greater Sage-Grouse* (Centrocercus urophasianus)

The greater sage-grouse is the largest grouse in North America (Mezquida et al. 2006). An obligate with sagebrush habitat, its current distribution is about half of its pre-settlement range (Schroeder et al. 2004), and many populations have been steadily declining in recent decades (Connelly and Braun 1997). Declines in sage-grouse in areas still dominated by sagebrush have been attributed to sagebrush removal through land conversion, nonnative plants (Wisdom et al. 2002), energy exploration and extraction (Braun ct al. 2002), grazing (Beck and Mitchell 2000), and altered fire regimes (Connelly et al. 2000). West Nile virus has also been a stressor (Naugle et al. 2004).

A recent climate change assessment for greater sage-grouse concluded that the cumulative effects of projected climate change on sagebrush and West Nile virus transmission would reduce suitable sage-grouse habitat in the Northern Rockies and northern Great Plains (Schrag et al. 2011). Because sage-grouse require large areas of mature sagebrush, future increases in wildfire are expected to significantly reduce habitat. Another assessment, focused on southeastern Oregon, concluded that in the near term, loss of sagebrush from wildfire and cheatgrass (*Bromus tectorum*) invasion will lead to habitat deterioration in future decades (Creutzburg et al. 2015). However, the same study also projected that native shrub-steppe communities would increase by around 2070, leading to habitat improvement.

8.3.15 Harlequin Duck (Histrionicus histrionicus)

Harlequin ducks in the intermountain West breed and summer on fast-flowing mountain streams and winter on rocky coastal areas. During summer, they feed primarily on larval insects on stream bottoms, and in winter on a variety of small food items including snails, small crabs, barnacles, and fish roe (Robertson and Goudie 2015). They are relatively rare in Montana, with a concentration in Upper McDonald Creek in Glacier National Park. In Glacier National Park, harlequin duck reproductive success declined with higher and less predictable streamflows (Hansen 2014), conditions that are expected to be more common in a warmer climate (Chap. 3).

8.3.16 Mountain Quail (Oreortyx pictus)

The mountain quail (*Oreortyx pictus*) is a small ground-dwelling bird that occupies upland forest and woodland habitats in the western United States and northern Mexico. In the northwestern United States, its range extends into deep canyons such as Hells Canyon of the Snake River, where populations have been declining (Pope and Crawford 2004). Population studies have focused on survival, but connections to climate-related change are minimal. Climate-related variables are, however, important to survival, with lower survival being linked to both hot, dry conditions and cold winter weather (Stephenson et al. 2011).

8.3.17 Pygmy Nuthatch (Sitta pygmaea)

The pygmy nuthatch, a tiny bird found throughout western North America, is a cavity nester, often associated with ponderosa pine forests, but also found in other forest types such as quaking aspen (*Populus tremuloides*). Pygmy nuthatches can exhibit a social structure of cooperative breeding in which "helpers" aid breeding birds by feeding the incubating female, feeding nestlings and fledglings, and defending nesting territory (Sydeman et al. 1988).

The pygmy nuthatch nests in cavities in both live and dead trees, and population responses to disturbance are modest. For example, population densities across a variety of thinning and fuel treatments remained constant except in thin-and-burn treatments, where densities increased over 500% (Hurteau et al. 2008). In another study, nuthatches showed a negative response one year after wildfire, but a neutral response in subsequent years (Saab et al. 2007). Given its apparent neutral response to disturbance, flexibility in habitat, and wide latitudinal range, it is difficult to project whether pygmy nuthatch will respond positively or negatively to climate change. Conversion of forest to non-forest can reduce habitat, but climate change is unlikely to cause significant population reductions.

8.3.18 Ruffed Grouse (Bonasa umbellus)

The ruffed grouse has a primarily boreal distribution that includes peninsular extensions into the Rocky Mountains and Appalachian Mountains. Ruffed grouse are commonly found in aspen forest, which provides important food sources (Stauffer and Peterson 1985), and use aspen stands of all ages (Mehls et al. 2014). Thus, optimal grouse habitat consists of aspen forests with stands in a variety of age classes, including a large component of young stands.

Aspen may be sensitive to heat and drought in some locations. Although higher temperature is expected to cause increased stress in aspen, differences in forest structure and age affect the relationship between aspen mortality and drought (Bell et al. 2014), and mortality can be reduced by controlling stand densities and ages and limiting competition from conifers. If climate change leads to decreased extent of aspen in the Northern Rockies, reduced habitat would have detrimental effects on ruffed grouse populations. Fortunately, good options exist to mitigate these changes through silviculture that favors aspen over conifers, and through active manipulation of stand densities and ages.

8.3.19 Columbia Spotted Frog (Rana luteiventris)

Columbia spotted frogs breed in montane ponds throughout western North America. The effects of climate change on these frogs are unclear. In Utah, they were more likely to occur in persistent, shady ponds that maintained constant temperatures (Welch and MacMahon 2005). In Yellowstone National Park, pond desiccation led to sharp declines in frog populations (McMenamina et al. 2008), and throughout their range, populations in large stable water bodies were healthy, whereas those in more ephemeral ponds were subject to rapid declines (Hossack et al. 2013). In

Montana, warmer winters were associated with improved reproduction and survival of Columbia spotted frogs (McCaffrey and Maxell 2010). This species does not appear to be sensitive to stand-replacing fires (Hossack and Corn 2007).

Columbia spotted frog populations are stable in areas with persistent water supplies, and are capable of rapid population expansion into restored wetlands (Hossack et al. 2013). However, the amphibian chytrid fungus (*Batrachochytrium dendrobatidis*, or Bd), is prevalent in many populations (Russell et al. 2010), and warmer water would favor Bd in most systems (see section on western toad below). Although the fungus is common, the population-level effects of infection are unclear.

8.3.20 Western Toad (Anaxyrus boreas)

The western toad is a montane amphibian broadly distributed across the western United States; in the southern Rocky Mountains, the subspecies boreal toad (*A. b. boreas*) is recognized. Western toads have declined in some locations, particularly at the southern extent of their range (Corn et al. 2005). This species suffers from Bd, which is often fatal. Laboratory studies of Bd have found that it grows optimally at 17–25 °C, and colonies are killed at 30 °C (Piotrowski et al. 2004). Although Bd can grow in temperatures as cold as 4 °C, warmer water would increase its prevalence. In a study across Colorado, Wyoming, and Montana, Bd was consistently found in western toad tissues, and was more prevalent in warmer, low-elevation sites (Muths et al. 2008). A warmer climate may allow Bd to spread to higher elevations and become even more widespread, although the susceptibility of western toads is uncertain, because increased mortality is not always associated with high infection rates.

8.4 Adapting Wildlife and Wildlife Management to Climate Change

Adaptation to climate change for wildlife resources in the Northern Rockies focused on maintaining adequate habitat and healthy wildlife populations, and increasing knowledge of species needs and climate sensitivities. In each workshop conducted by the Northern Rockies Adaptation Partnership (Chap. 1), participants identified major habitats in their subregion, then developed adaptation strategies for species they regarded as important and for which they believed viable management options exist. Here, we summarize adaptation options according to major habitats.

Riparian habitats are important across the Northern Rockies. The primary strategy for improving riparian habitat resilience is maintaining healthy American beaver populations. Beaver complexes can buffer riparian systems against low and high streamflows, and provide habitat structure and foraging opportunities for multiple species.

Quaking aspen habitats are common in the four western subregions and occur occasionally in the Grassland subregion. Aspen was identified as important because of its high productivity, role in structural diversity, and habitat for cavity-nesting birds. In the Greater Yellowstone Area, ruffed grouse was identified as strongly tied to aspen habitats. Reduced distribution and abundance of aspen is projected for some locations (especially lower elevation) in a warmer climate (Chap. 5). The most common tactics for promoting aspen resilience are allowing wildfire to burn or using prescribed fire in older aspen stands, protection from grazing, and reducing conifer encroachment in stands of any age.

Dry ponderosa pine forests are common in the Central Rockies and Eastern Rockies subregions, providing habitat for cavity-nesting birds such as the flammulated owl. These habitats have experienced encroachment by Douglas-fir as a result of fire exclusion, increasing vulnerability of pine to future fires. Tactics for promoting ponderosa pine resilience include reducing competition from Douglas-fir through understory burning and cutting, protecting mature stands, and planting ponderosa pine where it is no longer common.

The Western Rockies and Central Rockies subregions support *older, mesic forests* because they experience a maritime climate influence. These forests, which provide habitat for fisher, may have younger age classes (caused by increased disturbance; Chap. 7) and different species composition in a warmer climate (Chap. 5). Adaptation strategies include restoring historical structure, conserving current structure, and promoting potential future mesic forest habitats.

Mountain sagebrush-grassland habitat occurs in all subregions except the Grassland. In the Western Rockies subregion, these habitats have less of a sagebrush component, occur in steep mountain canyons, and support populations of mountain quail. In a warmer climate, these habitats could lose some of their forb component, making them vulnerable to increased abundance of nonnative species (Chap. 6). Specific tactics for restoring historical habitat and maintaining current habitat include assertively reducing sagebrush mortality from fire, controlling nonnative species, and restoring formerly cultivated lands.

Sagebrush habitats are common in the Eastern Rockies, Greater Yellowstone Area, and Grassland subregions, supporting greater sage-grouse, Brewer's sparrow, greater prairie chicken [*Tympanuchus cupido*], sharp-tailed grouse [*T. phasianellus*) and pygmy rabbit, among other species. Tactics for maintaining adequate sagebrush habitat include reducing sagebrush mortality from fire, controlling nonnative species, preventing fragmentation, and restoring degraded habitat. Current focus on conservation of greater sage-grouse within sagebrush habitat in the western United States will benefit from including a climate-smart approach to management.

In all subregions, independent of specific habitats, *a better understanding of species requirements and mechanisms of climate change effects* is needed. In addition, connectivity and the potential for disease may affect multiple habitats and species, although climate sensitivities of diseases are not well understood. There was wide agreement on the need to better understand the mechanisms of climate sensitivities relative to life histories of individual species. Examples of tactics to accomplish this objective include analyzing female Canada lynx home ranges to determine the necessary distribution and size of habitat patches, quantifying and monitoring pygmy rabbit distribution, and understanding sagebrush succession following fire. The influence of low-snow years on wolverine denning success is an example of a mechanistic relationship with climate that needs more data.

Connectivity was considered an important conservation strategy for most species in the Northern Rockies, although climate influences on connectivity are uncertain. Connectivity can be affected by changes in water supply, habitat loss, habitat shifts, vegetation phenology shifts, human population expansion and redistribution, and snowpack dynamics. Specific tactics that would improve maintenance of connectivity include monitoring with genetic, tracking, and remote-sensing tools; identifying dispersal habitats; and identifying and removing barriers to connectivity.

Disease is also important in most subregions, not tied to a particular habitat, and not well understood. Specific tactics for addressing disease include monitoring the presence of white-nose syndrome (caused by the fungus *Pseudogymnoascus destructans*) in bat hibernacula, monitoring disease trends in moose, and coordinating with state agencies to monitor West Nile virus.

References

Adams, R. A. (2010). Bat reproduction declines when conditions mimic climate change projections for western North America. *Ecology, 91,* 2437–2445.

Aubry, K. B., Raley, C. M., Buskirk, S. W., et al. (2013). Meta-analysis of habitat selection by fishers at resting sites in the Pacific coastal region. *Journal of Wildlife Management, 77,* 1937–2817.

Bagne, K. E., Friggens, M. M., & Finch, D. M. (Eds.). (2011). A system for assessing vulnerability of species (SAVS) to climate change (General Technical Report RMRS-GTR-257). Fort Collins: U.S. Forest Service, Rocky Mountain Research Station.

Barnett, T. P., Adam, J. C., & Lettenmaier, D. P. (2005). Potential impacts of a warming climate on water availability in snow-dominated regions. *Nature, 438,* 303–309.

Beck, J. L., & Mitchell, D. L. (2000). Influences of livestock grazing on sage grouse habitat. *Wildlife Society Bulletin, 28,* 993–1002.

Beever, E. A., Ray, C., Mote, P. W., & Wilkening, J. L. (2010). Testing alternative models of climate-mediated extirpations. *Ecological Applications, 20,* 164–178.

Beever, E. A., Ray, C., Wilkening, J. L., et al. (2011). Contemporary climate change alters the pace and drivers of extinction. *Global Change Biology, 17,* 2054–2070.

Beever, E. A., Dobrowski, S. Z., Long, J., et al. (2013). Understanding relationships among abundance, extirpation, and climate at ecoregional scales. *Ecology, 94,* 1563–1571.

Bell, D. M., Bradford, J. B., & Lauenroth, W. K. (2014). Forest stand structure, productivity, and age mediate climatic effects on aspen decline. *Ecology, 95,* 2040–2046.

Bergman, E. J., Bishop, C. J., Freddy, D. J., et al. (2014). Habitat management influences overwinter survival of mule deer fawns in Colorado. *Journal of Wildlife Management, 78,* 448–455.

Braun, C. E., Oedekoven, O. O., & Aldridge, C. L. (2002). Oil and gas development in western North America: Effects on sagebrush steppe avifauna with particular emphasis on sage grouse. *Transactions of the North American Wildlife and Natural Resources Conference, 67,* 337–349.

Campbell, R. D., Nouvellet, P., Newman, C., et al. (2012). The influence of mean climate trends and climate variance on beaver survival and recruitment dynamics. *Global Change Biology, 18*, 2730–2742.

Campbell, R. D., Newman, C., Macdonald, D. W., & Rosell, F. (2013). Proximate weather patterns and spring green-up phenology effect Eurasian beaver (*Castor fiber*) body mass and reproductive success: The implications of climate change and topography. *Global Change Biology, 19*, 1311–1324.

Case, M. J., Lawler, J. J., & Tomasevic, J. A. (2015). Relative sensitivity to climate change of species in northwestern North America. *Biological Conservation, 187*, 127–133.

Castillo, J. A., Epps, C. W., Davis, A. R., & Cushman, S. A. (2014). Landscape effects on gene flow for a climate-sensitive montane species, the American pika. *Molecular Ecology, 23*, 843–856.

Chalfoun, A. D., & Martin, T. E. (2010). Facultative nest patch shifts in response to nest predation risk in the Brewer's sparrow: A "win-stay, lose-switch" strategy? *Oecologia, 163*, 885–892.

Clark, D. R. (1988). Environmental contaminants and the management of bat populations in the United States. In R. C. Szaro, K. E. Severson, & D. R. Patton (Eds.), *Management of amphibians, reptiles, and small mammals in North America* (General Technical Report RM-166; pp. 409–413). Fort Collins: U.S. Forest Service, Rocky Mountain Research Station.

Connelly, J. W., & Braun, C. E. (1997). Long-term changes in sage grouse *Centrocercus urophasianus* populations in western North America. *Wildlife Biology, 3*, 229–234.

Connelly, J. W., Reese, K. P., Fischer, R. A., & Wakkinen, W. L. (2000). Response of sage-grouse breeding population to fire in southeastern Idaho. *Wildlife Society Bulletin, 28*, 90–96.

Copeland, J. P., McKelvey, K. S., Aubry, K. B., et al. (2010). The bioclimatic envelope of the wolverine: Do environmental constraints limit their geographic distribution? *Canadian Journal of Zoology, 88*, 233–246.

Corn, P. S., Hossack, B. R., & Muths, E. (2005). Status of amphibians on the continental divide: Surveys on a transect from Montana to Colorado, USA. *Alytes, 22*, 85–94.

Creutzburg, M. K., Henderson, E. B., & Conklin, D. R. (2015). Climate change and land management impact rangeland condition and sage-grouse habitat in southeastern Oregon. *AIMS Environmental Science, 2*, 203–236.

Davies, K. W., Boyd, C. S., Beck, J. L., et al. (2011). Saving the sagebrush sea: An ecosystem conservation plan for big sagebrush plant communities. *Biological Conservation, 144*, 2573–2584.

Edwards, A. C., Scalenghe, R., & Freppaz, M. (2007). Changes in the seasonal snow cover of alpine regions and its effect on soil processes: A review. *Quaternary International, 162*, 172–181.

Erb, L. P., Ray, C., & Guralnick, R. (2011). On the generality of a climate-mediated shift in the distribution of the American pika (*Ochotona princeps*). *Ecology, 92*, 1730–1735.

Fellers, G. M., & Pierson, E. D. (2002). Habitat use and foraging behavior of Townsend's big-eared bat (*Corynorhinus townsendii*) in coastal California. *Journal of Mammalogy, 83*, 167–177.

Foden, W. B., Butchart, S. H. M., Stuart, S. N., et al. (2013). Identifying the world's most climate change vulnerable species: A systematic trait-based assessment of all birds, amphibians and corals. *PloS One, 8*, e65427.

Foresman, K. R. (2012). *Mammals of Montana*. Missoula: Mountain Press Publishing Company.

Grayson, D. K. (2000). Mammalian responses to middle Holocene climatic change in the Great Basin of the Western United States. *Journal of Biogeography, 27*, 181–192.

Green, J. S., & Flinders, J. T. (1980). Habitat and dietary relationships of the pygmy rabbit. *Journal of Range Management, 33*, 136–142.

Gruver, J. C., & Keinath, D. A. (2003). Species assessment for Townsend's big-eared bat (*Corynorhinus townsendii*) in Wyoming. Cheyenne: Bureau of Land Management, Wyoming State Office.

Hansen, W. K. (2014). *Causes of annual reproductive variation and anthropogenic disturbance in harlequin ducks breeding in Glacier National Park, Montana*. Master's thesis. Missoula: University of Montana.

Henry, P., Sim, Z., & Russello, M. A. (2012). Genetic evidence for restricted dispersal along continuous altitudinal gradients in a climate change-sensitive mammal: The American pika. *PloS One, 7*, e39077.

Hodges, K. E. (2000). Ecology of snowshoe hares in northern boreal forests. In L. F. Ruggiero, K. B. Aubry, S. W. Buskirk, et al. (Eds.), *Ecology and conservation of lynx in the United States* (pp. 117–162). Boulder: University of Colorado Press.

Hossack, B. R., Adams, M. J., Pearl, C. A., et al. (2013). Roles of patch characteristics, drought frequency, and restoration in long-term trends of a widespread amphibian. *Conservation Biology, 27*, 1410–1420.

Hossack, B. R., & Corn, P. S. (2007). Responses of pond-breeding amphibians to wildfire: Short-term patterns in occupancy and colonization. *Ecological Applications, 17*, 1403–1410.

Hurteau, S. R., Sisk, T. D., Block, W. M., & Dickson, B. G. (2008). Fuel-reduction treatment effects on avian community structure and diversity. *Journal of Wildlife Management, 72*, 1168–1174.

Jarema, S. I., Samson, J., McGill, B. J., & Humphries, M. M. (2009). Variation in abundance across a species' range predicts climate change responses in the range interior will exceed those at the edge: A case study with North American beaver. *Global Change Biology, 15*, 508–522.

Katzner, T. E., & Parker, K. L. (1997). Vegetative characteristics and size of home ranges used by pygmy rabbits (*Brachylagus idahoensis*) during winter. *Journal of Mammalogy, 78*, 1063–1072.

Kie, J. G., Bowyer, R. T., Nicholson, M. C., et al. (2002). Landscape heterogeneity at differing scales: Effects on spatial distribution of mule deer. *Ecology, 83*, 530–544.

Krohn, W. B., Elowe, K. D., & Boone, R. B. (1995). Relations among fishers, snow, and martens: Development and evaluation of two hypotheses. *The Forestry Chronicle, 71*, 97–105.

Lankester, M. W. (2010). Understanding the impact of meningeal worm, *Parelaphostrongylus tenuis*, on moose populations. *Alces, 46*, 53–70.

Lenarz, M. S., Nelson, M. E., Schrage, M. W., & Edwards, A. J. (2009). Temperature mediated moose survival in northeastern Minnesota. *Journal of Wildlife Management, 73*, 503–510.

Linkhart, B. D., Reynolds, R. T., & Ryder, R. A. (1998). Home range and habitat of breeding flammulated owls in Colorado. *Wilson Bulletin, 110*, 342–351.

Lowe, S. J., Patterson, B. R., & Schaefer, J. A. (2010). Lack of behavioral responses of moose (*Alces alces*) to high ambient temperatures near the southern periphery of their range. *Canadian Journal of Zoology, 88*, 1032–1041.

Lucan, R. K., Weiser, M., & Hanak, V. (2013). Contrasting effects of climate change on the timing of reproduction and reproductive success of a temperate insectivorous bat. *Journal of Zoology, 290*, 151–159.

Luo, J. H., Koselj, K., Zsebok, S., et al. (2014). Global warming alters sound transmission: Differential impact on the prey detection ability of echolocating bats. *Journal of the Royal Society Interface, 11*, 20130961.

Magoun, A. J., & Copeland, J. P. (1998). Characteristics of wolverine reproductive den sites. *Journal of Wildlife Management, 62*, 1313–1320.

McCaffrey, R. M., & Maxell, B. A. (2010). Decreased winter severity increases viability of a montane frog population. *Proceedings of the National Academy of Sciences, USA, 107*, 8644–8649.

McKelvey, K. S., Aubry, K. B., & Ortega, Y. K. (2000). History and distribution of lynx in the contiguous United States. In L. F. Ruggiero, K. B. Aubry, S. W. Buskirk, et al. (Eds.), *Ecology and conservation of lynx in the United States* (pp. 207–264). Boulder: University of Colorado Press.

McKelvey, K. S., Copeland, J. P., Schwartz, M. K., et al. (2011). Climate change predicted to shift wolverine distributions, connectivity, and dispersal corridors. *Ecological Applications, 21*, 2882–2897.

McMenamina, S. K., Hadlya, E. A., & Wright, C. K. (2008). Climatic change and wetland desiccation cause amphibian decline in Yellowstone National Park. *Proceedings of the National Academy of Sciences, USA, 105*, 16988–16993.

Mehls, C. L., Jensen, K. C., Rumble, M. A., & Wimberly, M. C. (2014). Multi-scale habitat use of male ruffed grouse in the Black Hills National Forest. *The Prairie Naturalist, 46*, 21–33.

Mezquida, E. T., Slater, S. J., & Benkman, C. W. (2006). Sage-grouse and indirect interactions: Potential implications of coyote control on sage-grouse populations. *The Condor, 108*, 747–759.

Millar, C. I., & Westfall, R. D. (2010). Distribution and climatic relationships of the American pika (*Ochotona princeps*) in the Sierra Nevada and western Great Basin, U.S.A.: periglacial landforms as refugia in warming climates. *Arctic, Antarctic, and Alpine Research, 42*, 76–88.

Miller, R. F., & Eddleman, L. L. (2001). *Spatial and temporal changes of sage grouse habitat in the sagebrush biome* (Technical Bulletin 151). Corvallis: Oregon State University, Agricultural Experiment Station.

Mills, L. S., Zimova, M., Oyler, J., et al. (2013). Camouflage mismatch in seasonal coat color due to decreased snow duration. *Proceedings of the National Academy of Sciences, USA, 110*, 7360–7365.

Mote, P. W., Hamlet, A. F., Clark, M. P., & Lettenmaier, D. P. (2005). Declining mountain snowpack in western North America. *Bulletin of the American Meteorological Society, 86*, 39–49.

Mowat, G., Poole, K. G., & O'Donoghue, M. (2000). Ecology of lynx in northern Canada and Alaska. In L. F. Ruggiero, K. B. Aubry, S. W. Buskirk, G. M. Koehler, C. J. Krebs, M. K. KS, & J. R. Squires (Eds.), *Ecology and conservation of lynx in the United States* (pp. 265–306). Boulder: University Press of Colorado.

Murray, D. L., Cox, E. W., Ballard, W. B., et al. (2006). Pathogens, nutritional deficiency, and climate change influences on a declining moose population. *Wildlife Monographs, 166*, 1–30.

Murray, D. L., Hussey, K. F., Finnegan, L. A., et al. (2012). Assessment of the status and viability of a population of moose (*Alces alces*) at its southern range limit in Ontario. *Canadian Journal of Zoology, 90*, 422–434.

Muths, E., Pilliod, D. S., & Livo, L. J. (2008). Distribution and environmental limitations of an amphibian pathogen in the Rocky Mountains, USA. *Biological Conservation, 141*, 1484–1492.

Naugle, D. E., Aldridge, C. L., Walker, B. L., et al. (2004). West Nile virus: Pending crisis for greater sage-grouse. *Ecology Letters, 7*, 704–713.

Nichols, J. D., & Williams, B. K. (2006). Monitoring for conservation. *Trends in Ecology and Evolution, 21*, 668–673.

Noon, B. R., & Biles, C. M. (1990). Mathematical demography of spotted owls in the Pacific Northwest. *Journal of Wildlife Management, 54*, 18–27.

Norvell, R. E., Edwards, T. C., & Howe, F. P. (2014). Habitat management for surrogate species has mixed effects on non-target species in the sagebrush steppe. *Journal of Wildlife Management, 78*, 456–462.

Olson, L. E., Sauder, J. D., Albrecht, N. M., et al. (2014). Modeling the effects of dispersal and patch size on predicted fisher (*Pekania* [*Martes*] *pennanti*) distribution in the U.S. Rocky Mountains. *Biological Conservation, 169*, 89–98.

Perry, L. G., Andersen, D. C., Reynolds, L. V., et al. (2012). Vulnerability of riparian ecosystems to elevated CO_2 and climate change in arid and semiarid western North America. *Global Change Biology, 18*, 821–842.

Petersen, K. L., & Best, L. B. (1985). Brewer's sparrow nest-site characteristics in a sagebrush community. *Journal of Field Ornithology, 56*, 23–27.

Pierce, J. E., Larsen, R. T., Flinders, J. T., & Whiting, J. C. (2011). Fragmentation of sagebrush communities: Does an increase in habitat edge impact pygmy rabbits? *Animal Conservation, 14*, 314–321.

Piotrowski, J. S., Annis, S. L., & Longcore, J. E. (2004). Physiology of *Batrachochytrium dendrobatidis*, a Chytrid pathogen of amphibians. *Mycologia, 96*, 9–15.

Pope, M. D., & Crawford, J. A. (2004). Survival rates of translocated and native Mountain Quail in Oregon. *Western North American Naturalist, 64*, 331–337.

Purse, B. V., Mellor, P. S., Rodgers, D. J., et al. (2005). Climate change and the recent emergence of bluetongue in Europe. *Nature Reviews Microbiology, 3*, 171–181.

Raine, R. M. (1983). Winter habitat use and responses to snow cover of fisher (*Martes pennanti*) and marten (*Martes americana*) in southeastern Manitoba. *Canadian Journal of Zoology, 61*, 25–34.

Renecker, L. A., & Hudson, R. J. (1986). Seasonal energy expenditure and thermoregulatory response of moose. *Canadian Journal of Zoology, 64*, 322–327.

Robertson, G. J., & Goudie, R. I. (2015). Harlequin duck *Histrionicus histrionicus*. In A. Poole (Ed.), *The birds of North America online*. Ithaca: Cornell Laboratory of Ornithology. http://bna.birds.cornell.edu/bna. 11 Dec 2016.

Russell, D. M., Goldberg, C. S., Waits, L. P., & Rosenblum, E. B. (2010). *Batrachochytrium dendrobatidis* infection dynamics in the Columbia spotted frog *Rana luteiventris* in north Idaho, USA. *Diseases of Aquatic Organisms, 92*, 223–230.

Saab, V., Block, W., Russell, R., et al. (2007). *Birds and burns of the interior West: Descriptions, habitats, and management in western forests* (General Technical Report PNW-GTR-712). Portland: U.S. Forest Service, Pacific Northwest Research Station.

Sawyer, H., Nielson, R. M., Lindzey, F. G., et al. (2007). Habitat selection of Rocky Mountain elk in a nonforested environment. *Journal of Wildlife Management, 71*, 868–874.

Schrag, A., Konrad, S., Miller, S., et al. (2011). Climate-change impacts on sagebrush habitat and West Nile virus transmission risk and conservation implications for greater sage-grouse. *GeoJournal, 76*, 561–575.

Schroeder, M. A., Aldridge, C. L., Apa, A. D., et al. (2004). Distribution of sage-grouse in North America. *The Condor, 106*, 363–376.

Schwartz, M. K., Pilgrim, K. L., McKelvey, K. S., et al. (2004). Hybridization between Canada lynx and bobcats: Genetic results and management implications. *Conservation Genetics, 5*, 349–355.

Schwartz, M. K., Copeland, J. P., Anderson, N. J., et al. (2009). Wolverine gene flow across a narrow climatic niche. *Ecology, 90*, 3222–3232.

Schwartz, M. K., DeCesare, N. J., Jimenez, B. S., et al. (2013). Stand- and landscape-scale selection of large trees by fishers in the Rocky Mountains of Montana and Idaho. *Forest Ecology and Management, 305*, 103–111.

Sherwin, H. A., Montgomery, W. L., & Lundy, M. G. (2013). The impact and implications of climate change for bats. *Mammal Review, 43*, 171–182.

Smith, A. T., & Weston, M. L. (1990). Ochotona princeps. *Mammalian Species, 352*, 1–8.

Smucker, K. M., & Marks, J. S. (2013). Flammulated owls nest in hollow in ground. *Journal of Raptor Research, 47*, 421–422.

Squires, J. R., & Ruggiero, L. F. (2007). Winter prey selection of Canada lynx in northwestern Montana. *Journal of Wildlife Management, 71*, 310–315.

Squires, J. R., Decesare, N. J., Kolbe, J. A., & Ruggiero, L. F. (2008). Hierarchical den selection of Canada lynx in western Montana. *Journal of Wildlife Management, 72*, 1497–1506.

Squires, J. R., Decesare, N. J., Kolbe, J. A., & Ruggiero, L. F. (2010). Seasonal resource selection of Canada lynx in managed forests of the northern Rocky Mountains. *Journal of Wildlife Management, 74*, 1648–1660.

Squires, J. R., DeCesare, N. J., Olson, L. E., et al. (2013). Combining resource selection and movement behavior to predict corridors for Canada lynx at their southern range periphery. *Biological Conservation, 157*, 187–195.

Stauffer, F., & Peterson, S. R. (1985). Ruffed and blue grouse habitat use in southeastern Idaho. *Journal of Wildlife Management, 49*, 459–466.

Stephenson, J. A., Reese, K. P., Zager, P., et al. (2011). Factors influencing survival of native and translocated mountain quail in Idaho and Washington. *Journal of Wildlife Management, 75*, 1315–1323.

Sydeman, W. J., Güntert, M., & Balda, R. P. (1988). Annual reproductive yield in the cooperative pygmy nuthatch (*Sitta pygmaea*). *The Auk, 105*, 70–77.

Thorne, E. T., Williams, E. S., Spraker, T. R., et al. (1988). Bluetongue in free-ranging pronghorn antelope (*Antilocapra americana*) in Wyoming: 1976 and 1984. *Journal of Wildlife Diseases, 24*, 113–119.

U.S. Geological Survey (USGS). (2013). North American breeding bird survey data, 1996–2013. http://www.mbr-pwrc.usgs.gov/bbs/tr2013/tr05620.htm. 12 Dec 2016.

van Oort, H., McLellan, B. N., & Serrouya, R. (2011). Fragmentation, dispersal and metapopulation function in remnant populations of endangered mountain caribou. *Animal Conservation, 14*, 215–224.

Welch, N. E., & MacMahon, J. A. (2005). Identifying habitat variables important to the rare Columbia spotted frog in Utah (USA): an information-theoretic approach. *Conservation Biology, 19*, 473–481.

Wisdom, M. J., Rowland, M. M., Wales, B. C., et al. (2002). Modeled effects of sagebrush-steppe restoration on greater sage-grouse in the Interior Columbia Basin, USA. *Conservation Biology, 16*, 1223–1231.

Woods, B. A., Rachlow, J. L., Bunting, S. C., et al. (2013). Managing high-elevation sagebrush steppe: Do conifer encroachment and prescribed fire affect habitat for pygmy rabbits? *Rangeland Ecology Management, 66*, 462–471.

Yandow, L. (2013). Delineating limiting habitat features and climate variables for the American pika (*Ochotona princeps*). Master's thesis. Laramie: University of Wyoming.

Zimova, M., Mills, L. S., Lukacs, P. M., & Mitchell, M. S. (2014). Snowshoe hares display limited phenotypic plasticity to mismatch in seasonal camouflage. *Proceedings of the Royal Society of London B: Biological Sciences, 281*, 20140029.

Chapter 9
Effects of Climate Change on Recreation in the Northern Rockies

Michael S. Hand and Megan Lawson

Abstract Recreation has a significant economic impact throughout the Northern Rockies. A warmer climate will generally improve opportunities for warm-weather activities (hiking, camping, sightseeing) because it will create a longer time during which these activities are possible, especially in the spring and autumn "shoulder seasons." However, it will reduce opportunities for snow-based, winter activities (downhill skiing, cross-country skiing, snowmobiling) because snowpack is expected to decline significantly in the future. Recreationists will probably engage in more water-based activities in lakes and rivers in order to seek refuge from hotter summer weather. Higher temperatures may have both positive and negative effects on wildlife-based activities (hunting, fishing, birding) and gathering of forest products (e.g., berries, mushrooms), depending on how target habitats and species are affected.

Recreationists are expected to be highly adaptable to a warmer climate by shifting to different activities and different locations, behavior that is already observed from year to year. For example, downhill skiers may switch to ski areas that have more reliable snow, cross-country skiers will travel to higher elevations, and larger ski areas on federal lands may expand to multi-season operation. Water-based recreationists may adapt to climate change by choosing different sites that are less susceptible to changes in water levels. Hunters may need to adapt by altering the timing and location of hunts. Federal management of recreation is currently not very flexible with respect to altered temporal and spatial patterns of recreation. This can be at least partially resolved by assessing expected use patterns in a warmer climate, modifying opening times of facilities, and deploying seasonal employees responsible for recreational facilities earlier in the year.

M.S. Hand (✉)
U.S. Forest Service, Rocky Mountain Research Station, Washington, DC, USA
e-mail: mshand@fs.fed.us

M. Lawson
Headwaters Economics, Bozeman, MT, USA
e-mail: megan@headwaterseconomics.org

Keywords Recreation • Climate change • Adaptation • Northern Rockies • Skiing
• Hiking • Hunting • Fishing • Camping

9.1 Introduction

Lands administered by federal agencies and other organizations provide opportunities for outdoor recreation as an important benefit throughout the Northern Rockies. National forests in the U.S. Forest Service (USFS) Northern Region and Greater Yellowstone Area have 13.3 million visits per year; Yellowstone, Grand Teton, and Glacier National Parks have an additional 8 million visits per year (NPS 2014). Recreation opportunities in national forests and national parks are diverse in both type and location, with recreation experiences being largely inseparable from ecosystems and natural features. Natural and ecological conditions help determine the overall recreation experience, regardless of whether it consists of skiing, hiking, hunting, camping, visiting developed sites, exploring the backcountry, or simply a driving tour.

Climatic conditions and environmental characteristics determine the availability of and demand for recreation opportunities (Shaw and Loomis 2008). Changing climate may affect the supply of and demand for recreation opportunities, causing potential changes in visitation patterns, experiences, and benefits in the future. It has been suggested that climate change will increase outdoor recreation participation in general (Bowker et al. 2013), primarily because of increased summer and warm-weather activities outweighing decreased winter activities (Loomis and Crespi 2004; Mendelsohn and Markowski 2004). Variability can be expected both spatially and at seasonal and much longer time scales.

In a warming climate, federal land managers will face a complex and evolving challenge of managing recreation opportunities as ecological conditions and recreational preferences change. Investments in recreation infrastructure and facilities maintenance, and decisions about recreation development contribute to recreational setting and the kinds of recreational opportunities that are available. Federal agencies often classify these opportunities using the Recreation Opportunity Spectrum (ROS), which has been used in planning and management for decades (Clark and Stankey 1979). Recreation visitor behavior and values can be mapped in the ROS, providing managers with information about tradeoffs associated with different types of investments and development (Rosenthal and Walsh 1986; Swanson and Loomis 1996).

Although broad trends in recreation participation under climate change are expected, little is known about how recreation in the Northern Rockies will change. This chapter describes the broad categories of recreation activities that may be sensitive to climate-related changes in the Northern Rockies, using the available scientific literature to infer projected effects of climate change on recreation participation.

9.2 Relationships Between Climate Change and Recreation

Supply and demand for recreation opportunities are sensitive to climate via two general pathways: (1) *direct effects* of altered temperature and precipitation on availability and quality of recreation sites, and (2) *indirect effects* of climate on characteristics and ecological condition of recreation sites (Loomis and Crespi 2004; Mendelsohn and Markowski 2004; Shaw and Loomis 2008) (Fig. 9.1).

Direct effects of altered climate will affect most outdoor recreation activities in some manner, especially for skiing and other snow-based activities that depend on seasonal temperatures and the amount, timing, and availability of snow (Irland et al. 2001; Englin and Moeltner 2004; Stratus Consulting 2009). Warm-weather activities are also sensitive to direct effects of climate change. For example, higher minimum temperatures have been associated with increased national park visits in Canada, particularly during non-peak "shoulder" seasons (Scott et al. 2007). Number of warm-weather days is positively associated with expected visitation for national parks in the United States, although visitation may decline during extreme heat (Richardson and Loomis 2004; Bowker et al. 2012). Temperature and precipitation

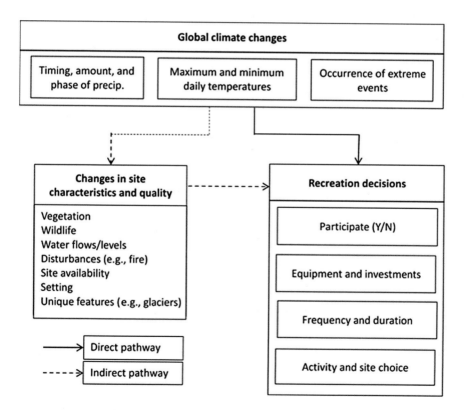

Fig. 9.1 Direct and indirect effects of climate on recreation decisions

will also affect the recreation experience (comfort, enjoyment) (Mendelsohn and Markowski 2004).

Indirect climate effects affect recreation activities that depend on additional ecosystem components, such as vegetation, surface water, and wildlife. Fishing for native cold-water species is expected to decrease as stream temperature increases, especially at lower elevation, where fish habitat will be degraded the most (Jones et al. 2013; Chap. 4). Surface water and streamflows are important for water-based recreation (e.g., boating), and forested area affects several outdoor activities (e.g., camping and hiking) (Loomis and Crespi 2004). Recreation visits to sites with valued natural characteristics (e.g., glaciers, charismatic wildlife species) (Chaps. 3, 8) or scenic qualities may decrease if the quality of those characteristics are threatened (Scott et al. 2007). Indirect effects of climate on disturbances, especially wildfire (Chap. 7), may also affect recreation behavior, although spatial and temporal patterns of recreation response will probably vary across the Northern Rockies (Englin et al. 2001).

9.3 Outdoor Recreation in the Northern Rockies

People participate in a wide variety of outdoor recreation activities in the Northern Rockies. The National Visitor Use Monitoring (NVUM) survey, administered by the USFS to monitor recreation visitation and activity in national forests, identifies 27 recreation activities in which people participate. Visitors are sampled using a stratified random sampling technique designed for assessing use on national forests. Sampling sites are stratified according to type of recreation site and times of day and week. Interviewees are selected at random, and are asked about different categories of travel-related spending within 80 km of the interview site (English et al. 2001). NVUM surveys include 25% of national forests in each region each year, so each unit is re-surveyed every 4 years. In this analysis, we used the most recent data available for each national forest, ranging from 2008 to 2012 (Fig. 9.2).

To assess how recreation patterns may change in the Northern Rockies, we identified categories of outdoor recreation expected to be sensitive to climate change. We defined a recreation activity as sensitive if changes in climate or climate-related conditions would potentially affect demand or supply for the activity. The 27 recreation activities in the NVUM survey were grouped into five climate-sensitive categories of activities, plus an "other" category of activities that are less sensitive to climate changes. Activities that comprise climate-sensitive categories are summarized in Table 9.1. These categories capture the most common types of recreation in public lands that would be affected by climate changes.

These 17 activities identified account for the primary recreation activities of 83% of visits to national forests in the Northern Rockies. Activities in the "Other" category are less sensitive to climate and are less frequently listed as a primary recreation activity. Warm-weather activities are the most popular (35.9% of visitors, 4.8 million per year), including hiking/walking, viewing natural features, developed

9 Effects of Climate Change on Recreation in the Northern Rockies

Fig. 9.2 Increased extent and severity of wildfires in a warmer climate will create forest conditions that may affect decisions by recreationists about hiking and other recreational activities (Photo courtesy of Dave Pahlas, http://IdahoAlpineZone.com)

and primitive camping, bicycling, backpacking, horseback riding, picnicking, and other non-motorized uses (Table 9.1). Hiking/walking was the most popular for 16.9% of visitors (2.2 million). Snow-based winter recreation was also popular (25% of visitors, 3.3 million), including downhill skiing, cross-country skiing, and snowmobiling. Wildlife-related activities, including hunting, fishing, and viewing wildlife, were the primary activity for 18.5% of visitors (2.5 million); hunting was the most popular with 11% of visitors (1.5 million). Gathering forest products, such as berries and mushrooms, was the primary activity for 2.4% of visitors (300,000). Motorized and non-motorized water activities (other than fishing) comprised 0.7% of visitors (97,000) (Fig. 9.3).

Non-local visitors (who live >50 km from the forest boundary) spent $601 million (in 2014 dollars) per year within 80 km of national forest boundaries (Table 9.2). This represents money spent in local communities that would not have occurred except for the motivation to recreate. Lodging expenses comprise 31% of total expenditures, followed by restaurants (18%), gas and oil (17%), and groceries (12%). Expenditures for other transportation, activities, admissions and fees, and souvenirs comprise 21% of spending.

Table 9.1 Participation in recreational activities in national forests in the Northern Rockies

Activity	Visitors for whom this was their primary activity — Number	Percent	Relationship to climate and environmental conditions
Warm-weather activities	**4,770,616**	**35.9**	Participation occurs during warm weather, dependent on the availability of snow- and ice-free sites, dry weather with moderate daytime temperatures, and availability of sites where air quality is not impaired by smoke from wildfires.
Hiking/walking	2,248,171	16.9	
Viewing natural features	1,136,245	8.6	
Developed camping	375,174	2.8	
Bicycling	286,707	2.2	
Other non-motorized	265,476	2.0	
Horseback riding	168,175	1.3	
Picnicking	164,638	1.2	
Primitive camping	74,876	0.6	
Backpacking	51,154	0.4	
Winter activities	**3,318,426**	**25.0**	Participation depends on timing and amount of precipitation as snow, and cold temperatures to support snow coverage. These activities are inherently sensitive to climatic variability and interannual weather patterns.
Downhill skiing	1,695,621	12.8	
Snowmobiling	843,778	6.4	
Cross-country skiing	779.027	5.9	
Wildlife activities	**2,452,053**	**18.5**	Temperature and precipitation are related to habitat suitability through effects on vegetation, productivity of food sources, species interactions, and water quantity and temperature (for aquatic species). Disturbances (wildland fire, invasive species, insect outbreaks) may affect amount, distribution, and spatial heterogeneity of suitable habitat.
Hunting	1,503,520	11.3	
Fishing	708,589	5.3	
Viewing wildlife	240,944	1.8	
Gathering forest products	**313,475**	**2.4**	Depends on availability and abundance of target species (e.g., berries, mushrooms), which are related to patterns of temperature, precipitation, and snowpack. Disturbances may alter availability and productivity of target species in current locations and affect opportunities for species dispersal.
Water-based activities (not including fishing)	**96,643**	**0.7**	Participation requires sufficient water flows (in streams and rivers) or levels (in lakes and reservoirs). Typically considered a warm-weather activity, and depends on moderate temperatures and snow- and ice-free sites. Some participants may seek water-based activities as a heat refuge during periods of extreme heat.

From USFS (n.d.)

Fig. 9.3 As snowpack decreases, opportunities for cross-country skiing at low elevation (shown in the Beaverhead-Deerlodge National Forest) may be available for shorter periods of time (Photo courtesy of U.S. Forest Service)

9.4 Assessing the Vulnerability of Recreation to Climate Change

The overall effect of climate change on recreation activity is likely to be an increase in participation and increase in the benefits derived from recreation. This is because warmer temperatures and increased season length will facilitate warm-weather activities, outweighing decreased winter activities that depend on snow and cold temperatures (Mendelsohn and Markowski 2004). However, these general findings obscure variation in recreation between types of activities and geographic locations. Here we assess the projected effects of climate on climate-sensitive recreation activities in the Northern Rockies, based on (1) reviews of existing studies of climate change effects on recreation and studies of how recreation behavior responds to

Table 9.2 Total annual expenditures by visitors to national forests in the Northern Rockies

Spending category	Non-local spending[a] Total annual expenditures *Thousands of $ (2014)*	Spending per category *Percent*	Local spending Total annual expenditures *Thousands of $ (2014)*	Spending per category *Percent*
Lodging	185,355	31	14,743	6
Restaurant	109,743	18	29,618	13
Groceries	74,003	12	44,886	19
Gasoline, oil	104,319	17	78,880	34
Other transportation	3013	1	1059	0
Activities	36,376	6	14,195	6
Admissions, fees	39,482	7	19,103	8
Souvenirs	48,839	8	28,075	12
Total	601,128		230,562	

From USFS (n.d.)
[a]Non-local refers to trips that required traveling more than 80 km

Fig. 9.4 Algal blooms (shown in Hayden Lake, Idaho) may be more common in a warmer climate, creating undesirable conditions for water-based recreation (Photo courtesy of Idaho Department of Environmental Quality)

climate-sensitive ecological characteristics, and (2) projections of climate-related biophysical changes described in other chapters in this volume (Fig. 9.4).

9.4.1 Current Conditions and Management

Public lands in the Northern Rockies provide an abundance and variety of recreational options, with opportunities for people of all interests and abilities. Opportunities range from high-use developed sites near urban areas and popular tourist destinations, to remote wilderness and seldom-used sites far from paved roads. Facilities and services vary; some sites are developed with modern amenities and staffed by agency employees, and others have little evidence of human influence other than a trailhead.

Wide variation in intra-annual and interannual weather and ecological conditions are normal, including broad variation in temperature, precipitation, water flows and levels, wildlife distributions, vegetative conditions, and wildfire activity. Most recreationists are already accustomed to making decisions about participating in recreation activities that incorporate a significant degree of uncertainty about conditions at different time scales (e.g., planning for a hike next week vs. planning for a ski trip next winter). Social factors, biogeographic conditions, and stressors all affect recreation in the Northern Rockies. Increased population, particularly in proximity to public lands, can strain visitor services and facilities due to increased use, and projected population increases in the future may exacerbate these effects (Bowker et al. 2012). Increased use caused by population growth can also reduce site quality because of congestion or damage at popular sites (Yen and Adamowicz 1994). Changes in land use may alter access to public lands, contribute to fragmentation of landscapes and habitat, and potentially alter disturbance regimes that affect recreation activities.

The physical condition of recreation sites and natural resources is dynamic, with variation caused by both human and natural forces. Recreation sites and physical assets need maintenance, and deferred or neglected maintenance may increase congestion at other sites that are less affected or increase hazards for visitors who continue to use degraded sites (USFS 2010). This stressor may interact with others, such as population growth and maintenance needs, if degraded site quality or congestion encourages users to engage in recreation that is not supported or appropriate at certain sites or at certain times of the year. Natural hazards and disturbances also affect recreation opportunities. For example, wildfire affects recreation demand as related to site quality and characteristics, but can also damage physical assets or exacerbate other natural hazards such as erosion (Chaps. 3, 11).

Recreation is an important component of the broader mission of public land management in the Northern Rockies. For lands managed by the USFS, sustainable recreation is a guiding principle for planning and management, seeking to "sustain and expand benefits to America that quality recreation opportunities provide" (USFS 2010, 2012b). Recreational resources are managed to connect people with natural resources and cultural heritage, and to adapt to changing social needs and environmental conditions.

9.4.2 Warm-Weather Activities

Warm-weather activities are the most common recreation activities in national forests and national parks in the Northern Rockies, comprising over one-third of all visits. Warm-weather recreation is sensitive to the length of appropriate season, depending on availability of snow- and ice-free trails and sites, and the timing and number of days having temperatures within minimum and maximum comfortable range (which may vary with activity type and site). The number of warm-weather days is a significant predictor of expected visitation behavior (Richardson and Loomis 2004), and studies of national park visitation show that minimum temperature is a strong predictor of monthly visitation patterns (Scott et al. 2007).

Participants are also sensitive to site quality and characteristics, such as the presence and abundance of wildflowers, trail conditions, and vegetation (e.g., cover for shade, wildfire effects). The condition of unique features that are sensitive to climate changes (e.g., glaciers) affects the desirability of certain sites (Scott et al. 2007). Forested area is positively associated with warm-weather activities, such as camping, backpacking, hiking, and picnicking (Loomis and Crespi 2004), and is sensitive to a warmer climate (USFS 2012a).

Wildfire can affect participation in warm-weather activities through changes to site quality and characteristics. The presence of burned forest areas can have different effects on the value of hiking trips (positive) and mountain biking (negative), although recent wildfire activity tends to decrease the number of visits (Englin et al. 2001; Loomis et al. 2001; Hesseln et al. 2003, 2004). High-severity fires are associated with decreased visitation, whereas low-intensity fires are associated with slight increases in visitation (Starbuck et al. 2006).

Recent fires are associated with initial losses of benefits for camping (Rausch et al. 2010) and backcountry recreation activities (Englin et al. 1996) that are attenuated over time. Visitation in Yellowstone National Park tends to be lower following months with high wildfire activity, although there is no discernable effect of previous-year fires (Duffield et al. 2013). Potential increases in the likelihood of extreme wildfire activity may reduce demand for warm-weather activities in certain years because of degraded site conditions, impaired air quality from smoke, and limited site access during and after fire management activities.

Demand for warm-weather activities is expected to increase because of a direct effect of warmer climate on season length, resulting in earlier availability of snow- and ice-free sites and an increase in the number of warm-weather days in spring and autumn. For example, higher minimum temperatures are associated with increased number of hiking days (Bowker et al. 2012). More extreme summer temperatures can dampen participation during the hottest weeks of the year, and extreme-heat scenarios for climate change are expected to reduce visitation (Richardson and Loomis 2004; Bowker et al. 2012). The temperature that is considered "extreme" may vary between individuals and chosen activities. Extreme heat may shift demand to cooler weeks at the beginning or end of the warm-weather season, or shift demand to sites that are less exposed to extreme temperature (e.g., higher elevations).

Adaptive capacity among recreationists is high because of the large number of potential alternative sites, ability to alter the timing of visits, and ability to alter capital investments (e.g., appropriate gear). However, benefits derived from recreation may decrease even if substitute activities or sites are available (Loomis and Crespi 2004). Access to alternative sites may involve higher costs (because of remoteness or difficulty of terrain) or higher congestion if demand is concentrated among fewer desirable locations. Although the ability of recreationists to substitute sites and activities is well established, it is unclear how people substitute across time periods or between large geographic regions (e.g., choosing a site in the Northern Rockies instead of the Northwest) (Shaw and Loomis 2008).

9.4.3 Cold-Weather Activities

Winter recreation sites in the Northern Rockies contain a wide range of characteristics, attracting local, national, and international visitors. Several sites support developed downhill skiing and snowboarding operated by special-use permit on lands administered by the USFS. Sites for cross-country skiing, snowshoeing, and snowmobiling are generally maintained directly by the USFS, and national parks also provide access for these activities.

Snow-based recreation is very sensitive to variations in temperature and the amount and timing of snow. Seasonal patterns of temperature and snowfall determine the viability and length of recreation seasons (Scott et al. 2008). Lower temperatures and the presence of new snow are associated with increased demand for skiing and snowboarding (Englin and Moeltner 2004). Indirect effects of climate, such as changes in scenery and unique features may affect winter recreation, but are expected to be relatively small.

Climate change will have negative effects on snow-based winter activities in the Northern Rockies, although effects will vary by site and especially by elevation. Warmer projected winter temperature for the region will reduce the proportion of precipitation as snow, even if total precipitation does not differ from historical amounts (Chap. 3). The rain-snow transition zone will move to higher elevations, particularly in late autumn and early spring (Klos et al. 2014), putting lower elevation sites at risk of shorter or non-existent winter recreation seasons. The highest elevation areas in the region are expected to remain snow-dominated through the end of the twenty-first century.

Studies of the ski industry in North America uniformly project negative effects of climate change (Scott and McBoyle 2007). Overall warming will reduce season length and likelihood of reliable winter recreation seasons. Climatological projections for the Northern Rockies (Chap. 2) are consistent with studies of ski area vulnerability to climate change in other regions, where projected effects of climate change on skiing, snowboarding, and other snow-based recreation activities is negative (Scott et al. 2008; Dawson et al. 2009; Stratus Consulting 2009).

Snow-based recreationists in the Northern Rockies have moderate capacity to adapt to changing conditions, because many winter recreation sites exist in the region. For minimally developed site activities (cross-country skiing, backcountry skiing, snowmobiling, snowshoeing), recreationists may seek higher elevation sites where snow is more likely to persist. Downhill skiing sites are fixed improvements, although potential adaptations include snowmaking, higher elevation development, and new run development (Scott and McBoyle 2007). Warmer temperature and higher precipitation as rain may increase availability of water for snowmaking in the near term, but warmer temperature also reduces the number of days when snowmaking is possible.

The Northern Rockies may have a comparative advantage in a warmer climate if the duration of snow-based recreation is longer than in other regions. In that case, recreationists may view recreation sites in the Northern Rockies as a substitute for other regions (e.g., the Southwest), although inter-regional substitution patterns are unclear (Shaw and Loomis 2008). Increased inter-regional substitution combined with shorter seasons may result in concentrated demand at fewer sites on fewer days, creating potential congestion.

9.4.4 Wildlife Activities

Wildlife recreation activities involve terrestrial or aquatic animals as a primary component of the recreation experience. Wildlife recreation can involve consumptive (e.g., hunting) or non-consumptive (e.g., wildlife viewing, birding, catch-and-release fishing) activities, and depend on distribution, abundance, and population of target species. These factors influence "catch rates," that is, the likelihood of catching or seeing an individual of the target species. Sites with higher catch rates can reduce costs associated with a wildlife activity (time and effort tracking targets), enhancing enjoyment of recreation (e.g., more views of a valued species).

Participation in wildlife activities is sensitive primarily to climate-related changes that affect expected catch rates. Catch rates determine site selection and trip frequency for hunting (Miller and Hay 1981; Loomis 1995), substitution among hunting sites (Yen and Adamowicz 1994), participation and site selection for fishing (Morey et al. 2002), and participation in non-consumptive wildlife recreation (Hay and McConnell 1979). Changes to habitat, food sources, or streamflows and water temperature (for aquatic species) may affect wildlife abundance and distribution, which in turn influences expected catch rates and recreation behavior.

The availability of highly valued targets affects benefits derived from wildlife activities (e.g., cutthroat trout [*Oncorhynchus clarkii*] for cold-water anglers) (Pitts et al. 2012), as does species diversity for hunting (Milon and Clemmons 1991) and non-consumptive activities (Hay and McConnell 1979). Temperature and precipitation are related to general trends in participation (Mendelsohn and Markowski 2004; Bowker et al. 2012), although the precise relationship may be specific to activity and species. For example, hunting for deer and elk (*Cervus elaphus*) is enhanced by cold temperatures and snowfall to aid in tracking, field dressing, and packing out animals.

Warmer temperature is expected to increase participation in terrestrial wildlife activities in the Northern Rockies because more days will be available for recreation. This concurs with previous documentation that hunting, birding, and viewing wildlife are associated with warmer weather (Bowker et al. 2012). However, hunting that occurs during discrete seasons may depend on weather conditions during a short period of time, and desirability of hunting may decrease if warmer weather reduces snow cover at specific times.

Habitat for target species is a function of interactions among species dynamics, vegetation, and disturbances, making it difficult to project the effects of climate change on habitat in complex landscapes. Although vegetative productivity may decrease in the future, effects on hunted species populations may be neutral, depending on size, composition, and spatial heterogeneity of vegetation used as forage (Chap. 8). The effects of disturbances on target species harvest rates will vary considerably depending spatial and temporal patterns of post-disturbance vegetation and on animal species requirements.

Higher temperatures will decrease populations of native cold-water fish species as climate refugia retreat to higher elevations (Chap. 4), with potential increases in (nonnative) fish species that can tolerate warmer temperatures. However, it is unclear if shifting populations of species (e.g., substituting rainbow trout [*O. mykiss*] for cutthroat trout) will affect catch rates because relative abundance of fish may not change. Higher interannual variability in precipitation, extreme drought, and reduced snowpack could lead to higher peakflows in winter and lower low flows in summer, creating stress for fish populations. Increased incidence and severity of wildfire may increase the likelihood of erosion that degrades aquatic habitat, degrading the quality of individual streams and potentially reducing the desirability of angling as compared to other activities.

9.4.5 Gathering Forest Products

Gathering forest products for recreational purposes accounts for a small portion of primary recreational activities in the Northern Rockies, although it is more common as a secondary activity (e.g., as part of day hike). Forest products are also important for cultural and spiritual uses. An avid population of enthusiasts for certain types of products supports a small but steady demand for gathering activities. Small-scale commercial gathering competes with recreationists for popular and high-value products such as huckleberries (*Vaccinium* spp.) in some locations.

Forest product gathering is sensitive to climatic and vegetative conditions that support the distribution and abundance of target species. Participation is comparable to warm-weather recreation, depending on moderate temperatures and accessibility of target sites. Vegetative change and increased interannual variation in precipitation may alter the geographic distribution and productivity of target species

(Chaps. 5, 6). Increased wildfires may eliminate sources of forest products in some locations (e.g., berries), but in some cases may encourage short- or medium-term productivity for other products (e.g., mushrooms). Long-term changes that reduce forest cover may decrease viability of gathering in areas that transition to less productive vegetation.

Recreationists engaged in forest product gathering may be able to select different gathering sites as the distribution and abundance of target species changes, although tradeoffs may exist, such as increased travel and expense. Those who engage in gathering as a secondary activity can select alternate activities to complement primary activities. The magnitude of climate effects on forest product gathering is expected to be low—it is generally not a primary activity, and users may be able to substitute other sites or activities without much loss in recreational value. Longer warm-weather seasons may increase opportunities for gathering, although these changes may not correspond with greater availability of target species. The likelihood of effects is expected to be moderate, although significant uncertainty exists regarding direct and indirect effects on forest product gathering.

9.4.6 Water-Based Activities (Not Including Fishing)

Apart from angling, water-based activities are a small portion of primary recreation activity participation on federal lands. Lakes and reservoirs provide opportunities for both motorized and non-motorized boating and swimming, although boating is commonly paired with fishing. Existing stressors include the occurrence of drought conditions that reduce water levels and site desirability in some years, and disturbances that can alter water quality (e.g., erosion following wildfires).

Availability of desirable locations for water-based recreation is sensitive to reduced water levels caused by warming temperatures, increased variability in precipitation (including severe droughts), and decreased precipitation as snow. Lower water levels may also have an indirect effect on the aesthetic qualities of some water-based recreation sites (e.g., exposure of "bathtub rings" at reservoirs with low water levels). Reduced surface-water area is associated with less participation in boating and swimming (Loomis and Crespi 2004; Mendelsohn and Markowski 2004; Bowker et al. 2012), and streamflow is positively associated with number of days spent rafting, canoeing, and kayaking (Loomis and Crespi 2004). Warmer temperature is also generally associated with higher participation in water-based activities (Loomis and Crespi 2004; Mendelsohn and Markowski 2004), although extreme heat may reduce participation (Bowker et al. 2012).

Increasing temperatures, reduced storage of water as snowpack, and increased variability of precipitation are expected to increase the likelihood of reduced water levels and greater variation in lake levels on federal lands (Chap. 3), which is associated with reduced site quality and suitability for some activities. Increased demand for surface water by downstream users may exacerbate low water levels in drought years. Warmer temperatures are expected to increase demand for water-based recreation as the viable season lengthens, and although extreme heat encourages some people to seek water-based activities to cool off, it can also discourage participation in outdoor recreation in general (Bowker et al. 2012). Overall, projections of water-based activities in response to climate change tend to be small compared to the effects of broad population and economic shifts (Bowker et al. 2012).

9.4.7 Summary

Several recreation activities in the Northern Rockies are considered sensitive to direct effects of a warmer climate and indirect effects on site conditions and extreme events (including disturbances) (Table 9.3). However, recreation activities are diverse, and the effects of climate will vary widely between different activities and across geographic areas within the region. Overall, participation in recreation activities is expected to increase, primarily because longer warm-weather seasons will

Table 9.3 Summary of climate change assessment for recreation in the Northern Rockies, where positive (+) and negative (−) signs indicate expected direction of effect on overall benefits derived from recreation activity

Activity	Magnitude of climate effect	Likelihood of climate effect	Direct effects	Indirect effects
Warm-weather activities	Moderate (+)	High	Warmer temperature (+), higher likelihood of extreme temperatures (−)	Increased incidence, area, and severity of wildfire (+/−), increased smoke from wildfire (−)
Snow-based winter activities	High (−)	High	Warmer temperature (−), reduced precipitation as snow (−)	
Wildlife activities	Terrestrial wildlife: low (+); fishing: moderate (−)	Moderate	Warmer temperature (+), higher incidence of low streamflow (fishing -), reduced snowpack (hunting -)	Increased incidence, area, and severity of wildfire (terrestrial wildlife +/−), reduced cold-water habitat, incursion of warm-water tolerant species (fishing -)
Gathering forest products	Low (+/−)	Moderate	Warmer temperature (+)	More frequent wildfires (+/−), higher severity wildfires (−)
Water-based activities (not including fishing)	Moderate (+)	Moderate	Warming temperatures (+), higher likelihood of extreme temperatures (−)	Lower streamflows and reservoir levels (−)

make more recreation sites available for longer periods of time. Increased participation in warm-weather activities will probably be offset to some extent by decreased snow-based activities. Receding snow and shorter seasons in the future will reduce the number of available days and sites for winter recreation.

Recreation demand is governed by several economic decisions with interacting dependencies on climate. For example, decisions on whether to engage in winter recreation, activity type (e.g., downhill vs. cross-country skiing), location, frequency of participation, and duration of stay per trip depend on climatic and ecological characteristics of recreation sites. Climatic effects on recreation depend on spatial and temporal relationships between sites, as well as on biophysical conditions, and human decisions.

The exact effects of climate on recreation sites and target species will be difficult to predict across the Northern Rockies, although these effects will play a role in recreation decisions for some activities. The adaptation response of recreationists is also uncertain, because inter-regional and inter-temporal substitution behavior is poorly understood (Shaw and Loomis 2008), although substitution will almost certainly be an important consideration. Many popular activities have several alternate sites, or timing of visits can be altered in response to a warmer climate. However, substitution may lead to reduced benefits if the alternate sites are more difficult or costly to access, or provide a lower quality recreational experience.

9.5 Adapting Recreation and Recreation Management to Climate Change

9.5.1 Adaptation by Recreation Participants

For the most part, warm-weather recreationists will benefit from a warmer climate through a longer recreation season, and will not need to adapt significantly. If extreme heat becomes more common, they have the option of selecting alternate sites at higher elevation or perhaps near water. Increasing temperatures will have significant negative effects on snow-based recreation, reducing season length and possibly snow quality. Many skiers will have the option of going to other ski areas (downhill skiing) or sno-parks (cross-country skiing) that have suitable snow, requiring them to be aware of local conditions and often be willing to travel farther.

Water-based recreationists may adapt to climate change by choosing different sites that are less susceptible to changes in water levels (e.g., by seeking higher-elevation natural lakes) and changing the type of water-based recreation activity (e.g., from motorized boating on reservoirs to non-motorized boating on natural lakes).

Hunters may need to adapt by altering the timing and location of hunts. However, state rules on hunting season dates impose a constraint on this behavior unless states change hunting seasons based on expected climate changes. Hunters may also target different species if the abundance or distribution of preferred species changes.

Wildlife viewers may change the timing and location of viewing experiences and target different species. They have more flexibility than hunters to shift timing to coincide with appropriate weather conditions or species movements. Anglers may adapt by choosing different species to target (e.g., shifting from cold-water to warm-water species) and choosing sites that are less affected by higher temperatures (e.g., higher-elevation streams). The former is less costly, although some anglers may place a high value on certain target species and have a lower willingness to target warm-water species.

9.5.2 Adaptation by Federal Land Management

Resource managers may need to reconsider how infrastructure investments and availability of facilities align with changing ecological conditions and demands for recreation settings. The ROS can be used to match changing conditions and preferences with available opportunities. Adaptation may include responding to changing recreation patterns, but also helping to shape the settings and experiences available to recreationists on public lands.

Recreation managers have options for responding to changing patterns in warm-season recreation demand in order to provide sustainable recreation opportunities. A critical first step will be to assess changing patterns of use, then adjust as necessary to increase capacity of recreation sites with higher use (e.g., campgrounds can be enlarged, more signs and gates can be installed). Some adjustments may be driven by increased congestion and resource damage, although expansion may be limited in locations that have environmental constraints (e.g., USFWS 2013). Timing of trail closures, food storage orders, and special-use permits may need to be adjusted to ensure sustainable recreation programs. For example, the season for whitewater rafting permits may need to be modified in response to altered streamflows.

A general adaptation strategy for winter recreation is to transition recreation management to address shorter winter recreation seasons and changing recreation use. There may be opportunities to expand facilities into areas of concentrated use. In addition, snow-based recreation can be diversified to include more snowmaking, additional ski lifts, and higher-elevation runs. Adaptation tactics related to supply and quality of winter recreation could result in tradeoffs with other activities, including warm-weather access to high-elevation sites or effects of snowmaking on streamflow.

Increased frequency of disturbances (fire, flooding) have the potential to cause increased damage to infrastructure associated with recreation activities. Recreation sites can be managed to decrease risks to public safety and infrastructure. Assessments can be used to determine which sites and infrastructure are most at risk from disturbance, and strategic investments can be made to ensure that facilities are sustainable in the future and accommodate changing use.

References

Bowker, J. W., Askew, A. E., Cordell, H. K., et al. (2012). *Outdoor recreation participation in the United States—Projections to 2060: A technical document supporting the Forest Service 2010 RPA assessment* (General Technical Report GTR-SRS-160). Asheville: U.S Forest Service, Southern Research Station.

Bowker, J. M., Askew, A. E., Poudyal, N. C., et al. (2013). Climate change and outdoor recreation participation in the Southern United States. In J. M. Vose & K. D. Klepzig (Eds.), *Climate change adaptation and mitigation management options: A guide for natural resource managers in the Southern forest ecosystems* (pp. 421–450). Boca Raton: CRC Press.

Clark, R. N., & Stankey, G. H. (1979). The *recreation opportunity spectrum: A framework for planning, management, and research* (General Technical Report PNW-GTR-098). Portland: U.S. Forest Service, Pacific Northwest Research Station.

Dawson, J., Scott, D., & McBoyle, G. (2009). Climate change analogue analysis of ski tourism in the northeastern USA. *Climate Research, 39*, 1–9.

Duffield, J. W., Neher, C. J., Patterson, D. A., & Deskins, A. M. (2013). Effects of wildfire on national park visitation and the regional economy: A natural experiment in the Northern Rockies. *International Journal of Wildland Fire, 22*, 1155–1166.

Englin, J., & Moeltner, K. (2004). The value of snowfall to skiers and boarders. *Environmental and Resource Economics, 29*, 123–136.

Englin, J., Boxall, P. C., Chakraborty, K., & Watson, D. O. (1996). Valuing the impacts of forest fires on backcountry forest recreation. *Forest Science, 42*, 450–455.

Englin, J., Loomis, J., & González-Cabán, A. (2001). The dynamic path of recreational values following a forest fire: A comparative analysis of states in the Intermountain West. *Canadian Journal of Forest Research, 31*, 1837–1844.

English, D., Kocis, S., Zarnoch, S., & Arnold, R. J. (2001). *Forest Service National Visitor Use Monitoring process: Research method documentation* (General Technical Report SRS-GTR-57). Athens: U.S. Forest Service, Southern Research Station.

Hay, M. J., & McConnell, K. E. (1979). An analysis of participation in nonconsumptive wildlife recreation. *Land Economics, 55*, 460–471.

Hesseln, H., Loomis, J. B., González-Cabán, A., & Alexander, S. (2003). Wildfire effects on hiking and biking demand in New Mexico: A travel cost study. *Journal of Environmental Management, 69*, 359–368.

Hesseln, H., Loomis, J. B., & González-Cabán, A. (2004). The effects of fire on recreation demand in Montana. *Western Journal of Applied Forestry, 19*, 47–53.

Irland, L. C., Adams, D., Alig, R., et al. (2001). Assessing socioeconomic impacts of climate change on U.S. forests, wood-product markets, and forest recreation. *Bioscience, 51*, 753–764.

Jones, R., Travers, C., Rodgers, C., et al. (2013). Climate change impacts on freshwater recreational fishing in the United States. *Mitigation and Adaptation Strategies for Global Change, 18*, 731–758.

Klos, P. Z., Link, T. E., & Abatzoglou, J. T. (2014). Extent of the rain-snow transition zone in the western U.S. under historic and projected climate. *Geophysical Research Letters, 41*, 4560–4568.

Loomis, J. (1995). Four models for determining environmental quality effects on recreational demand and regional economics. *Ecological Economics, 12*, 55–65.

Loomis, J., & Crespi, J. (2004). Estimated effects of climate change on selected outdoor recreation activities in the United States. In R. Mendelsohn & J. Neumann (Eds.), *The impact of climate change on the United States economy* (pp. 289–314). Cambridge: Cambridge University Press.

Loomis, J., González-Cabán, A., & Englin, J. (2001). Testing for differential effects of forest fires on hiking and mountain biking demand and benefits. *Journal of Agricultural and Resource Economics, 26*, 508–522.

Mendelsohn, R., & Markowski, M. (2004). The impact of climate change on outdoor recreation. In R. Mendelsohn & J. Neumann (Eds.), *The impact of climate change on the United States economy* (pp. 267–288). Cambridge: Cambridge University Press.

Miller, J. R., & Hay, M. J. (1981). Determinants of hunter participation: Duck hunting in the Mississippi flyway. *American Journal of Agricultural Economics, 63*, 401–412.

Milon, J. W., & Clemmons, R. (1991). Hunters' demand for species variety. *Land Economics, 67*, 401–412.

Morey, E. R., Breffle, W. S., Rowe, R. D., & Waldman, D. M. (2002). Estimating recreational trout fishing damages in Montana's Clark Fork river basin: Summary of a natural resource damage assessment. *Journal of Environmental Management, 66*, 159–170.

National Park Service (NPS). (2014). *National park service visitor use statistics, park reports, 5-year visitation summary, 2010–2014*. https://irma.nps.gov/Stats/Reports/Park. 26 June 2015.

Pitts, H. M., Thacher, J. A., Champ, P. A., & Berrens, R. P. (2012). A hedonic price analysis of the outfitter market for trout fishing in the Rocky Mountain West. *Human Dimensions of Wildlife, 17*, 446–462.

Rausch, M., Boxall, P. C., & Verbyla, A. P. (2010). The development of fire-induced damage functions for forest recreation activity in Alberta, Canada. *International Journal of Wildland Fire, 19*, 63–74.

Richardson, R. B., & Loomis, J. B. (2004). Adaptive recreation planning and climate change: A contingent visitation approach. *Ecological Economics, 50*, 83–99.

Rosenthal, D. H., & Walsh, R. G. (1986). Hiking values and the recreation opportunity spectrum. *Forest Science, 32*, 405–415.

Scott, D., & McBoyle, G. (2007). Climate change adaptation in the ski industry. *Mitigation and Adaptation Strategies for Global Change, 12*, 1411–1431.

Scott, D., Jones, B., & Konopek, J. (2007). Implications of climate and environmental change for nature-based tourism in the Canadian Rocky Mountains: A case study of Waterton Lakes National Park. *Tourism Management, 28*, 570–579.

Scott, D., Dawson, J., & Jones, B. (2008). Climate change vulnerability of the U.S. Northeast winter recreation-tourism sector. *Mitigation and Adaptation Strategies for Global Change, 13*, 577–596.

Shaw, D., & Loomis, J. (2008). Frameworks for analyzing the economic effects of climate change on outdoor recreation and selected estimates. *Climate Research, 36*, 259–269.

Starbuck, C. M., Berrens, R. P., & McKee, M. (2006). Simulating changes in forest recreation demand and associated economic impacts due to fire and fuels management activities. *Forest Policy and Economics, 8*, 52–66.

Stratus Consulting. (2009). *Climate change in Park City: an assessment of climate, snowpack, and economic impacts* (Report prepared for The Park City Foundation). Washington, DC: Stratus Consulting, Inc. http://www.parkcitymountain.com/site/mountain-info/learn/environment. 23 Mar 2015.

Swanson, C. S., & Loomis, J. B. (1996). *Role of nonmarket economic values in benefit-cost analysis of public forest management options* (General Technical Report PNW-GTR-361). Portland: U.S. Department of Agriculture, Forest Service, Pacific Northwest Research Station. 32p.

U.S. Fish & Wildlife Service (USFWS). (2013). *Northern Continental Divide ecosystem grizzly bear conservation strategy (draft)*. http://www.fws.gov/mountain-prairie/species/mammals/grizzly/NCDE_Draft_CS_Apr2013_Final_Version_corrected_headers.pdf. 24 Aug 2015.

U.S. Forest Service (USFS). (2010). Connecting people with America's great outdoors: A framework for sustainable recreation. http://www.fs.usda.gov/Internet/FSE_DOCUMENTS/stelprdb5346549.pdf. 24 March 2015.

U.S. Forest Service (USFS). (2012a). *Future of America's forest and rangelands: Forest Service 2010 Resources Planning Act assessment* (General Technical Report WO-87). Washington, DC: U.S. Forest Service.

U.S. Forest Service (USFS). (2012b). National forest system land management planning, 36 CFR Part 219, RIN 0596-AD02. *Federal Register, 77*, 21162–21276.

U.S. Forest Service (USFS). (n.d.). *Calculations of National Visitor Use Monitoring survey data, round 2 (Custer, Bridger-Teton, Gallatin, Shoshone National Forests) and round 3 (Beaverhead-Deerlodge, Caribou-Targhee National Forests)*. http://www.fs.fed.us/recreation/programs/nvum. 24 Mar 2015.

Yen, S. T., & Adamowicz, W. L. (1994). Participation, trip frequency and site choice: A multinomial-Poisson hurdle model of recreation demand. *Canadian Journal of Agricultural Economics, 42*, 65–76.

Chapter 10
Effects of Climate Change on Ecosystem Services in the Northern Rockies

Travis Warziniack, Megan Lawson, and S. Karen Dante-Wood

Abstract Ecosystem services are increasingly valued on federal lands, beyond just their economic value. Climate change effects will vary greatly within different subregions of the Northern Rockies, with some ecosystem services being affected in the short term and others in the long term. Of the many ecosystem services provided in the Northern Rockies, eight are considered here, including annual water yield, water quality, wood products, minerals and mineral extraction, forage for livestock, viewsheds and air quality, regulation of soil erosion, and carbon sequestration.

Although annual water yield is not expected to change significantly, timing of water availability will likely shift, and summer flows may decline. These changes may result in some communities experiencing summer water shortages, although reservoir storage can provide some capacity. Rural agricultural communities will be disproportionately affected by climate change if water does become limiting. Water quality will also decrease in some locations if wildfires and floods increase, adding sediment to rivers and reservoirs. Hazardous fuel treatments, riparian restoration, and upgrading of hydrologic infrastructure can build resilience to disturbances that damage water quality.

Forage for livestock is expected to increase in productive grasslands as a result of a longer growing season and in some cases elevated carbon dioxide. Therefore, ranching and grazing may benefit from climate change. Primary effects on grazing include loss of rural population, spread of nonnative grasses, and fragmentation of rangelands.

Viewsheds and air quality will be negatively affected by increasing wildfires and longer pollen seasons. A growing percentage of the Northern Rockies population

---

T. Warziniack (✉)
U.S. Forest Service, Rocky Mountain Research Station, Fort Collins, CO, USA
e-mail: twwarziniack@fs.fed.us

M. Lawson
Headwaters Economics, Bozeman, MT, USA
e-mail: megan@headwaterseconomics.org

S. Karen Dante-Wood
U.S. Forest Service, Office of Sustainability and Climate, Washington, DC, USA
e-mail: skdante@fs.fed.us

will be in demographic groups at risk for respiratory and other medical problems on days with poor air quality. Hazardous fuel treatments can help build resilience to disturbances that degrade air quality.

Carbon sequestration will be increasingly difficult if wildfires, insect outbreaks, and perhaps plant disease increase as expected, especially in the western part of the Northern Rockies. At the same time, managing forests for carbon sequestration is likely to become more important in response to national policies on carbon emissions. Hazardous fuel treatments can help build resilience to disturbances that rapidly oxidize carbon and emit it to the atmosphere.

Keywords Water yield • Water quality • Wood products • Minerals • Viewsheds • Air quality • Soil erosion • Carbon sequestration • Climate change • Adaptation • Ecosystem services • Social vulnerability • Rocky Mountains • Natural capital

10.1 Introduction

Ecosystem services are benefits to people from the natural environment, including timber for wood products, clean water for downstream users, recreation opportunities, and spiritual and cultural connection to the environment and natural resources. As stated by Collins and Larry (2007), "An ecosystem services perspective encourages natural resource managers to extend the classification of multiple uses to include a broader array of services or values."

Categorizing ecosystem services (Box 10.1) helps identify ways in which natural resources benefit humans, and how changes in the natural environment will affect these benefits. These categories are not exclusive, and many natural resources fall

---

Box 10.1 Ecosystem Services Definitions
From the Millennium Ecosystem Assessment (2005)

Provisioning services: Products obtained from ecosystems, including timber, fresh water, wild foods, and wild game.

Regulating services: Benefits from the regulation of ecosystem processes, including the purification of water and air, carbon sequestration, and climate regulation.

Cultural services: Nonmaterial benefits from ecosystems, including spiritual and religious values, recreation, aesthetic values, and traditional knowledge systems.

Supporting services: Long-term processes that underlie the production of all other ecosystem services, including soil formation, photosynthesis, water cycling, and nutrient cycling.

under multiple categories. For example, *consumption of water* is a provisioning service, the process of *purifying water* a regulating service, *recreational use of water* a cultural service, and the role of *water in the life history of animals* a supporting service. Climate change will affect the quality and quantity of ecosystem services (positively and negatively) provided by public lands. Establishing the link among natural processes, ecosystem services, and human benefits helps clarify the communities or types of people most vulnerable to a changing climate.

Lands in the Northern Rockies provide ecosystem services to people who visit, live adjacent to, or otherwise benefit from natural resources on public lands. First, we introduce ecosystem services and how to describe and measure them. Second, we describe how people and communities use and benefit from public lands, as well as existing stressors that may affect the ability of communities to adapt to a changing climate. Third, we discuss climate change effects on specific ecosystem services. Finally, we identify adaptation strategies that can help reduce negative effects, and discuss the adaptive capacity of agencies and communities.

10.2 Ecosystem Services on Public Lands in the Northern Rockies

There are many beneficiaries from ecosystem services provided by public lands, including neighboring communities, non-local visitors, and people who may never visit or directly use the lands but gain satisfaction from knowing a resource exists (Kline and Mazzotta 2012). This is particularly true for iconic landscapes and rivers in the study area (e.g., Yellowstone National Park; Borrie et al. 2002). Managing for multiple use of natural resources can create situations in which some ecosystem services conflict with others. For example, managing for non-motorized recreation may conflict with managing for motorized recreation, timber, and mining, but it could complement management for biodiversity and some wildlife species.

Ecosystem services from public lands are critical for neighboring communities, particularly in rural areas of the Northern Rockies where people rely on these lands for fuel, food, water, recreation, and cultural connection. Decreased quantity and quality of ecosystem services produced by public lands will affect human systems that rely on them, requiring some communities to seek alternative means of providing services or to change local economies and lifeways.

Management decisions for public lands can affect ecosystem service flows, with cascading effects on numerous users. We will highlight climate change effects on ecosystem services flows, and how management decisions can help users mitigate or adapt to these effects. The concept of ecosystem services is relatively new, so data on this topic are scarce. Although we use quantitative data when possible, we often rely on qualitative descriptions or proxy measures. Demographic and economic factors provide an important context for understanding the effects of climate change on ecosystem services.

> **Box 10.2 Ecosystem Services Assessed in the Northern Rockies**
>
> **Provisioning Ecosystem Services**
>
> - Abundant fresh water for human (e.g., municipal and agricultural water supplies) and environmental (e.g., maintaining streamflow) uses
> - Building materials and wood products
> - Mining materials
> - Forage for livestock
> - Fuel from firewood and biofuels
> - Air quality and scenic views
> - Genetic diversity and biodiversity
>
> **Regulating Ecosystem Services**
>
> - Water filtration and maintenance of water quality associated with drinking, recreation, and aesthetics
> - Protection from wildfire and floods
> - Protection from erosion
> - Carbon sequestration

We focus on provisioning and regulating ecosystem services in the Northern Rockies (Box 10.2). The amount of detail for these ecosystem services varies, depending on how much information is available and can be interpreted in the context of climate change. Several of the ecosystem services are also discussed in other chapters, including genetic diversity and biodiversity (Chap. 5), protection from wildfire and floods (Chap. 7), and recreation (Chap. 9). Ecosystem services are combined in a single section if all of them are likely to be affected by the same changes in natural resource conditions.

10.3 Social Vulnerability and Adaptive Capacity

Social vulnerability analyses seek to identify which institutions, resources, and characteristics make communities more or less resilient to environmental hazards. The most widely used measure of social vulnerability is the Social Vulnerability Index (SoVI) (Cutter et al. 2003), which is based on 11 factors: personal wealth, age, density of the built environment, single-sector economic dependence, housing stock and tenancy, race (separate factors for African American and Asian), ethnicity (separate factors for Hispanic and Native American), occupation, and infrastructure dependence. Scores based on these factors are summed to form a composite vulnerability score for each county in the United States.

Most counties in the Northern Rockies are in the high to medium vulnerability range, which is typical for areas dominated by rural economies. The average percentage of county populations living in rural areas in the Northern Rockies is 75.3%, compared to a national average of 19.3% (based on the 2012 Census American Community Survey). Rural counties tend to rely on a single industry, have older populations, and have fewer social resources (e.g., hospitals) than urban areas. The oldest mean age in the region is in Prairie County, Montana, where the mean age is 56, and the average median age among the counties is 43.4. An aging population and decline in youth in rural counties worries many because of the potential loss of a traditional culture in many Western communities.

The median household income in the Northern Rockies is $45,235, considerably lower than the national average of $53,046. High-income counties tend to be in the eastern part of the region with ties to the oil and gas industry, and areas with recreation-based businesses; low-income counties often depend on grazing and timber. Unemployment and poverty are relatively widespread in the region (Fig. 10.1), although the region as a whole had an average unemployment rate in 2012 of 5.4%, lower than the national average of 9.3%. Spatially, unemployment follows median incomes closely, with counties in the east having low unemployment and counties in the west having high unemployment. Many of the factors that make individuals more vulnerable are compounded among migrants and minorities. They tend to have fewer economic resources, lack political power, and sometimes struggle with communication (Fothergill and Peek 2004).

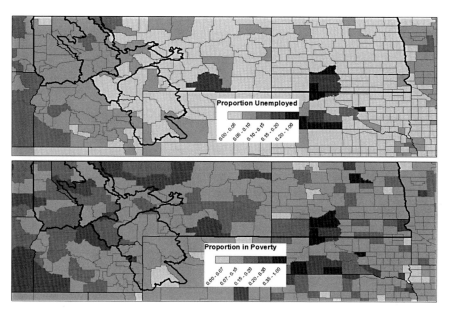

Fig. 10.1 Demographic information for unemployment (upper) and poverty (lower) in the Northern Rockies

10.4 Assessing the Effects of Climate Change on Ecosystem Services

10.4.1 Water Quantity

Major consumptive uses of water in the Northern Rockies include domestic and municipal water supply, industrial use of water, and water for oil and gas development (drilling and hydraulic fracturing). Non-consumptive uses of water include recreational uses (e.g., boating, maintaining fish habitat) and hydroelectric power production. Most water in the Northern Rockies is already appropriated, and many uses are tied to junior water rights that can be exercised only during high-flow years. Any new uses of water require a transfer of water rights, increased water supply through reservoir storage, or mining of ground water.

A recent draft of the Montana State Water Plan details water use in Montana (Tables 10.1 and 10.2) and is representative of most of the Northern Rockies. Hydroelectric power generation (hydropower) accounts for 86% of total water

Table 10.1 Total water use in Montana

Planning basin	Hydropower	Irrigation	Reservoir evaporation	Municipal, industrial, livestock	Instream flow
			Percent		
Statewide	85.9	12.4	1.2	0.5	0
Clark Fork/Kootenai River	94.4	4.7	0.5	0.4	0
Upper Missouri	88.0	11.2	0.5	0.3	0
Lower Missouri	39.2	19.5	6.0	0.3	35.0
Yellowstone River	24.5	23.0	0.4	1.4	50.7

From Montana DNRC (2014)

Table 10.2 Consumptive water use in Montana

Planning basin	Irrigation	Reservoir evaporation	Domestic & municipal	Livestock	Industrial	Thermo-electric
			Percent			
Statewide	67.3	28.0	2.4	1.2	0.3	0.8
Clark Fork/Kootenai River	66.4	27.0	3.9	0.5	1.2	0
Upper Missouri	82.3	13.7	3.0	0.9	<0.1	0
Lower Missouri	42.0	56.2	0.4	1.4	<0.1	0
Yellowstone River	83.4	7.2	2.8	2.1	0.3	4.2

From Montana DNRC (2014)

demand in Montana (Montana DNRC 2014), although hydropower is considered a non-consumptive use because it does not affect instream flow or total water downstream. However, reservoirs needed for hydropower experience high rates of water loss to evaporation. Fort Peck Reservoir, in the Lower Missouri River Basin, annually loses 754,000 megaliters of water to evaporation. The largest consumptive use of water in Montana is irrigated agriculture, which accounts for 96% of all water diversions and 67% of all consumptive use. The marginal value of water in agriculture is an order of magnitude lower than the marginal value of water for municipal uses (Montana DNRC 2014).

Compared to more arid regions of the western United States, changes in water yield in the Northern Rockies are expected to be modest, although they may be disproportionately large for local residents who experience them (Foti et al. 2012). Climate and hydrologic models consistently project changes in timing of runoff, making the likelihood of these effects high. Warmer temperatures will make drought more frequent, despite small increases in precipitation shown in some climate models, increasing overall competition for water. This will amplify many of the effects of population growth and demographic changes already occurring. Agricultural and municipal users will experience major impacts, making it more difficult to allocate instream flows for recreation and wildlife.

Timing of snowmelt is a major concern in the Columbia and Missouri Basin headwaters. Earlier runoff may be out of sync with many of the demands for water by agriculture, even as warmer months extend the growing season. Future water quantities in North Dakota and the eastern plains of Montana are likely to be more variable. Higher temperatures have already brought a mixture of impacts to agriculture in North Dakota, where wheat production alone generates $4.5 billion annually in economic activity (North Dakota Wheat Commission 2007). Warmer temperatures and higher commodity prices have pushed wheat and corn production into areas of the state where either they were not previously grown or where shorter-season varieties dominated.

Drier soils and more intense precipitation events may increase flood frequency, leading to increased dependence on tile drainage. In 2002, drought cost North Dakota $223 million. In 2005, heavy rains ruined 400,000 hectares of cropland and prevented another 400,000 from being planted, causing $425 million in damage (Karetinkov et al. 2008). More droughts and intense temperatures may also make plants more susceptible to insect pests (Rosenzweig et al. 2000). More frequent droughts and heavy rain events will stress municipal water supply systems and infrastructure.

Climate change will make it harder to preserve instream flows in the future, with small mountain streams and valued fisheries being particularly vulnerable. Some of the most productive waterfowl breeding grounds in the northern prairie wetlands and pothole region (> 50% of North America's ducks breed here) will be threatened in a warmer climate, and unless these wetlands are maintained, bird populations will be significantly reduced (Sorenson et al. 1998; Johnson et al. 2005).

Transfer of water rights from one use to another is legally possible within the Northern Rockies but realistically constrained by the ability to transport water.

Transfers between agricultural and municipal uses, for example, can occur only between users in the same watershed. Because municipal values of water are so high, these transfers are likely to occur if demand is high enough.

Re-use of effluent and other conservation methods will be important tools for adaptation. Groundwater pumping is a short-term solution, but is not sustainable in the long run. These methods are expensive and will be cost prohibitive for most rural communities in the Northern Rockies. New municipal demands are more likely to be met by purchasing or leasing reliable senior water rights (Montana DNRC 2014). Water rights are still available in some water basins, but they are junior in priority and not reliable for municipal uses. A drier climate in prairie pothole habitats of the Grassland subregion will make it desirable to preserve and restore waterfowl habitat along the wetter fringes (Johnson et al. 2005).

10.4.2 Water Quality, Aquatic Habitats, and Fish

Headwater streams in the Northern Rockies generally provide safe, clean drinking water to downstream communities, and water is important to cultural practices of Native Americans, including the ability to exercise their fishing rights. However, many of the region's streams and lakes are already threatened or impaired according to U.S. Environmental Protection Agency standards, with impairment being caused by grazing, feedlots, and fertilizer runoff. Runoff from roads and bridges are a problem in Idaho, leading to high levels of phosphorous and mercury.

Disturbances such as wildfires and mudslides are a major concern for municipal water supplies (Chap. 3). Sudden increases in sediment or other pollutants often cause water treatment plants to shut down or incur high costs to remove sediment from reservoirs. Some Northern Rockies residents worry about the effects of increased oil and gas extraction activities on watershed health. Groundwater contamination in northeastern Montana near the Fort Peck Indian Reservation has been linked to development of the East Poplar oil field (Thamke and Smith 2014). Oil spills in the Yellowstone River (2011, 2015) and a pipeline leak near Tioga, North Dakota (2014) highlight the dangers to watersheds surrounding oil and gas fields.

Climate change will potentially affect fishing, water-based recreation, and drinking water, amplifying the effects of development on water quality already occurring in the region. Increased number and severity of wildfires will deposit more sediment in streams, lakes, and reservoirs. Increased air temperature and loss of vegetation along stream banks will raise the temperature of streams, and altered vegetation may affect water filtration and flow rate.

Warming air temperature will cause stream temperatures to increase. Some native fish species, such as bull trout (*Salvelinus confluentus*) and cutthroat trout (*Oncorhynchus clarkii*), are extremely sensitive to warm water, whereas some non-native species can tolerate higher temperatures (Chap. 4). Fish habitats at lower elevations are likely to experience the biggest, near-term temperature increases, making them vulnerable to shifts in species composition and distribution. Native fish species with high ecological plasticity will be able to withstand some environmental change by altering life history timing or distribution patterns, but the magnitude and rate of change will overwhelm species with narrow ecological niches (Chap. 4). Culturally important fisheries, such as those of the Nez Perce tribe, will be affected within the boundaries of their reservation and traditional fishing grounds, already stressed by hydropower and stream modification (Wagner et al. 2004).

Warming has already led to expansion of agriculture in some areas of the Northern Rockies. Continued expansion will generally decrease water quality, but the net effects of flooding and drought are uncertain (Warziniack 2014). Lower water flows have also been linked to increased water temperature, eutrophication, and content of nutrients and metals (Murdoch et al. 2000; van Vliet et al. 2011), especially in nutrient-rich bodies of water (Schindler et al. 2008).

Restoration of streams, wetlands, and riparian areas may help stabilize water temperatures in some locations, but in the long term, investments in water treatment infrastructure will be needed if sediment increases substantially or if large disturbances become more frequent. Enhancing fish populations through hatcheries is already occurring, and such human intervention may become more important in the future. Other adaptation strategies are described in Chaps. 4 and 9.

10.4.3 Building Materials and Wood Products

Timber and forest products are dominant economic forces in the Northern Rockies, with forest products comprising 23% of direct manufacturing employment in Montana (McIver et al. 2013) (Table 10.3). Because much of the timber in the Northern Rockies is exported from the region, the most important aspect of timber is providing jobs, particularly in rural communities. The timber industry also provides a labor force capable of doing forest restoration work. In 2011, Idaho and Montana contained 160 timber processing facilities including 73 saw mills. Over 97% of timber is processed in sawmills, up from 80% (Keegan et al. 2005).

Historically, much of the timber in the area has come from national forests, although that share has decreased greatly. In 1979, 46% of timber harvested in Idaho came from national forests, declining to only 7% in 2006 (Brandt et al. 2012). Timber removal has varied over time in response to changing market and policy conditions, but the past decade has been particularly difficult for the timber industry (Table 10.3). Between 2005 and 2009, employment in the wood products industry declined 29% in Idaho and 24% in Montana (Keegan et al. 2012). Mills in the region are the major employer for some small communities, making the effects of mill closures particularly pronounced in a few places. Although timber jobs have been declining in the Northern Rockies, non-timber jobs have been increasing.

The direct effect of climate on timber production is expected to be small. More important to the timber industry are the economic and policy changes that affect demand for forest products and timber quotas for national forests. The primary sensitivities of timber resources associated with climate change are wildfire, insects,

Table 10.3 Sold timber volume from national forests in the U.S. Forest Service Northern Region and Greater Yellowstone Area subregion

	1980			2013		
National Forest	Sales	Sold volume Thousands of m^3	Inflation adjusted sold value US dollars	Sales	Sold volume Thousands of m^3	Inflation adjusted sold value US dollars
Beaverhead-Deerlodge	630	111	1,971,012	845	19	59,067
Bitterroot	268	101	3,883,685	266	19	459,684
Bridger-Teton	425	48	885,087	627	23	150,834
Caribou-Targhee	7347	232	7,726,627	743	17	93,922
Custer	127	4	81,794	292	4	18,088
Flathead	289	459	22,504,836	334	35	963,163
Gallatin	310	56	628,518	551	11	44,820
Helena	113	52	1,451,979	393	8	34,000
Idaho Panhandle	669	748	64,207,103	866	95	3,562,340
Kootenai	616	415	36,705,744	820	84	1,820,020
Lewis and Clark	277	29	134,615	387	5	21,160
Lolo	367	96	2,281,829	597	15	298,537
Nez Perce-Clearwater	414	603	18,881,743	699	105	6,567,655
Shoshone	307	28	198,089	415	18	225,075

Data from U.S. Forest Service, via Headwater Economics; http://headwaterseconomics.org/interactive/national-forests-timber-cut-sold

and disease (Ryan et al. 2008; Chaps. 6, 7). In addition, warmer winters and associated freezing and thawing may increase forest road erosion and landslides, making winter harvest more difficult and expensive. Climate change will result in larger, more frequent fires and a longer fire season. Increased fires may increase demand for fuel treatments, either through timber harvests or through mechanical and manual thinning that uses the timber labor force and infrastructure. Although this may affect the availability of harvestable wood products, the overall effect on timber-related jobs would be relatively small.

Management actions may be able to mitigate drought stress and soil water deficits. Land managers also have the option to conduct fuel treatments, which help decrease the probability of large, severe wildfires and to salvage burned or insect-killed timber before it loses market value. However, timber management cannot respond quickly to potential threats. The wood products industry may be able to adapt to changing conditions by using alternative species, changing the nature or location of capital and machinery, changing reliance on imports or exports, and adopting new technologies (Irland et al. 2001). The most resilient communities will be those that diversify their economic bases, effectively reducing their exposure to adverse impacts to the timber industry.

10.4.4 Mining Materials

Minerals are provisioning ecosystem services, but their primary role in the region is as an economic driver, providing jobs and incomes. Mineral development is important throughout the Northern Rockies, but particularly in northeastern Montana and northwestern North Dakota. In some counties, oil and gas development represents a third of total income to residents. The main stressors from oil and gas development are effects on other ecosystem services, such as water quality. Traffic from trucks and heavy machinery also increase the risk of introducing nonnative species to surrounding rangelands.

Climate will not directly affect minerals industries, although power generation, oil and gas development, and mineral extraction are major users of water. Increased mudslides and fires may threaten oil and gas infrastructure, which would in turn threaten the ecosystem services that are co-located with mineral development. Regional centers of oil and gas draw people from all over the country looking for high-paying jobs. Competition for workers in the oil fields causes wages in all other sectors of regional economics, including traditionally low-wage jobs in the service industry, to rise. If climate adversely affects other economic sectors, job opportunities in mining and energy will become more important. Climate change could affect the oil and gas infrastructure, but non-climatic drivers will be more important, including international prices for oil and gas, national climate policy, and regional concerns about threats to watersheds.

Global economic forces primarily drive the oil and gas industry. Oil and gas development potential determines where drilling activity takes place, and regional growth occurs so quickly that communities respond to, rather than plan for, such development. The most successful mineral-based economies are those that are able to collect some of the resource rents from drilling and invest them in the community, extending prospects for long-term economic growth (Kunce and Shogren 2005).

10.4.5 Forage for Livestock

The area contained within the Northern Rockies Adaptation Partnership contains 64 million hectares of rangeland, of which 85% are privately held. Of the federal rangeland, 3.4 million hectares are Bureau of Land Management lands, of which 3.2 million hectares are in Montana. Most counties in the region have a significant share of total income derived from cattle.

Cheatgrass (*Bromus tectorum*) and other nonnative plants have become a major nuisance in Western rangelands, significantly reducing usable forage. Human modification has also converted rangeland to other uses, dominated by agricultural development, resource extraction, and residential development (Reeves and Mitchell 2012). Human modification and fragmentation of rangelands have potential consequences for socio-economic sustainability of rural communities, including loss of rural character, loss of biodiversity, difficulty in managing interconnected lands for grazing, threats to watershed health, compromised viewscapes, loss of native species, and changes in disturbance regimes.

Warmer temperatures carbon dioxide fertilization are expected to increase productivity of rangelands (Reeves and Mitchell 2012; Chap. 6), and increased regional population will lead to fragmentation of rangelands. Arid grasslands are likely to experience short-term response in species richness because of the prevalence of annual species (Cleland et al. 2013). Carbon dioxide enrichment may alter the relative abundance of grassland plant species by increasing the production of one or more species without affecting biomass of other dominant species (Polley et al. 2012).

Cattle stocking rates in the Northern Rockies remain at or below current capacity of the land to support livestock (Reeves and Mitchell 2012), with few counties experiencing forage demand above current forage supply. The biggest threat to rangeland from climate change may be increasing rates of spread of nonnative weeds and changes in fire regime (Maher 2007). Fire makes ranch planning difficult. Loss of access to grazing areas requires emergency measures like the use of hay, requiring increased investments by ranchers. Increased fire will facilitate conversion of more lands to domination by nonnative plants. Fire also kills shrubs, especially sagebrush (*Artemisia* spp.), increasing the prevalence of grasses and herbs, thus reducing structural and floristic diversity.

Human modification of rangelands and associated fragmentation are driven by opportunities for economic growth, as land is converted to higher value uses. Rangeland conversion to residential development has brought new populations, higher incomes, and higher tax bases to rural communities, creating what has been called the "New West" (Riebsame et al. 1997). Natural amenities in and near the Rocky Mountains are often touted as an economic asset (Power 1998; Rasker 1993), and during the 1990s, 67% of counties in the Rocky Mountains grew faster than the national average (Beyers and Nelson 2000). The effects of demographic and socioeconomic factors may affect rangeland quantity and quality more than climate change in some areas.

10.4.6 Viewsheds and Clean Air

Air quality is an ecosystem service that can be altered by changes in vegetation composition and tree responses to climate change. Tropospheric ozone, air pollution episodes, plant sensitivity to air pollutants, and release of pollen all affect the provision of air quality by forests. The Northern Rockies generally have exceptional air quality, although a few counties in the region regularly have days with poor air quality (American Lung Association 2015), and some areas are subject to wintertime inversions that trap air pollutants. During inversions, wood-burning stoves become a major source of air pollution. In summer, smoke from wildfires settles in valleys, leading to poor air quality. Some areas in Idaho are affected by burning of crop residues, and smoke can get trapped or settle into valleys where it persists until strong winds clear the air. Major sources of air pollution in North Dakota include coal-fired power plants, oil-field emissions, and vehicle traffic in mineral-rich areas of the state.

A large percentage of Northern Rockies residents are in demographic groups that are sensitive to poor air quality (e.g., elderly, poor), and nearly 1 in 10 adults have asthma (Center for Disease Control 2009). As more young people leave rural communities, sensitive populations remain in rural areas without health facilities that can accommodate an aging population.

Air quality can deteriorate rapidly during a wildfire, and increased frequency of wildfires will affect viewsheds and air quality. Extended fire seasons will affect both scenery and air quality, with detrimental effects to human health (Bedsworth 2011). Climate change may affect distribution patterns and mixtures of air pollutants through altered wind patterns and amount and intensity of precipitation. By 2050, summertime organic aerosol concentration over the western United States is projected to increase by 40% and elemental carbon by 20%. Higher temperatures accelerate chemical reactions that, in combination with reactive hydrocarbons, form ozone and secondary particles (Kinney 2008).

Systems are already in place to alert residents when air quality deteriorates. Adaptation options include limiting physical activity outdoors, using air conditioning, and taking medications to mitigate health impacts. Tighter restrictions on use of wood for heating homes and on agricultural burning can reduce pollutants, and fuel treatments can reduce wildfire risk and smoke production. As noted above, the effects of poor air quality are greater for vulnerable populations like the elderly, young, and poor, who have little capacity to adapt.

10.4.7 Regulation of Soil Erosion

A U.S. Forest Service (USFS) soil management directive identifies six soil functions: soil biology, soil hydrology, nutrient cycling, carbon storage, soil stability and support, and filtering and buffering (USFS 2009). Steep slopes are the key element associated with erosion and landslides in mountain landscapes, and open rangeland is susceptible to topsoil loss. Erosion and landslides threaten infrastructure, water quality, and important cultural sites. Resource management practices are designed to limit erosion and soil compaction, but landslides and erosion are still a common problem, with roads and other human activities serving as a large source of sediment in many watersheds. Loss of soil from agricultural fields is a problem in the eastern part of the Northern Rockies, but best practices in agriculture and range management have begun to slow the loss. Soil loss rates still exceed natural regeneration of soil in much of this area, and may continue with further expansion of agriculture.

Soil erosion interacts with other landscape processes affected by climate change. In mountainous areas, wildfire and precipitation interact to affect erosion rates. Frequency of wildfire, precipitation in the form of rain rather than snow, and intense precipitation events may lead to greater erosion and more landslides. A combination of increased drought and flooding could exacerbate erosion in some areas. Erosion is also a significant concern for cultural sites (Chap. 11).

One of the key impacts of soil erosion in mountains is its effect on water quality and water treatment costs. Without expensive dredging, the usable life of dams and reservoirs will decrease, and capital investments will be necessary to remove sediment (Sham et al. 2013). Limiting erosion on rangelands can be done with best management practices for agriculture, including the use of buffers and limiting activity in sensitive riparian areas. Accelerating fuel treatments to make forests more resilient to wildfire can limit erosion by reducing fire severity.

10.4.8 Carbon Sequestration

Forests provide an important ecosystem service in the form of carbon sequestration—the uptake and storage of carbon in forests and wood products. Carbon sequestration is considered a regulating ecosystem service because it mitigates greenhouse gas emissions by offsetting losses through removal and storage of carbon. Carbon storage in forests is becoming more valuable as the impacts of greenhouse gas emissions manifest in different ways (USFS 2015).

The National Forest System includes 22% of the total U.S. forest area and 24% of total carbon stored in U.S. forests (excluding interior Alaska). Carbon sequestration can be enhanced by preventing conversion of forest land to non-forest uses, restoring and maintaining resilient forests better adapted to a changing climate and other stressors, and reforesting lands disturbed by wildfires. Federal agencies balance carbon stewardship with a wide range of ecosystem services by maintaining and enhancing net sequestration in existing ecosystems.

Although disturbances are the predominant drivers of forest carbon dynamics, biogeochemical cycling and climatic variability influence forest growth rates and consequently the carbon fluxes (Pan et al. 2009, 2011). Changes in carbon stocks and resulting net emissions may be influenced by vegetation management and restoration—fire and fuels management, timber harvest, reforestation, and other practices—that can integrate carbon dynamics across broad landscapes and over many decades, while meeting other resource management objectives. Harvested wood products (HWP), such as lumber, panels, and paper, can account for a significant amount of off-site carbon storage, contributing to national-level accounting and regional reporting (Skog 2008). Products derived from timber harvest from federal lands extend carbon storage and/or substitute for fossil fuel use.

Estimates of total ecosystem carbon and stock change (flux) have been produced for all national forests in the United States (USFS 2015). Carbon stocks reflect the amount of carbon stored in aboveground live trees, belowground live trees, understory, standing dead trees, down dead wood, forest floor, and soil organic carbon. Carbon stock change (flux) reflects year-to-year balance of carbon (Woodall et al. 2013) and measures interannual variation caused by tree growth, disturbance, and management.

Carbon stock trends for each national forest between the years 1990 and 2013 (Fig. 10.2) indicate that Idaho Panhandle National Forest stored the largest amount of carbon in the region (188 Tg in 1990, 183 Tg in 2013), and some forests had an increase in carbon stocks and others a decrease. Cumulative carbon stored in HWP in the USFS Northern Region increased in 1955, peaking in 1995 with 34 Tg (Fig. 10.3). Since then, the HWP pool has decreased to 32 Tg, illustrating how timber harvest affects HWP.

Many factors affect the sensitivity of forests to sequester carbon, and although the net effect of climate change on carbon storage in forests is uncertain, increased risk of wildfire and insect outbreaks will make it more difficult to retain carbon in biomass. Preliminary results from the Forest Carbon Management Framework (Healey et al. 2014; Raymond et al. 2015) show that fire had a major impact on carbon storage in Flathead National Forest between 1990 and 2012, followed by

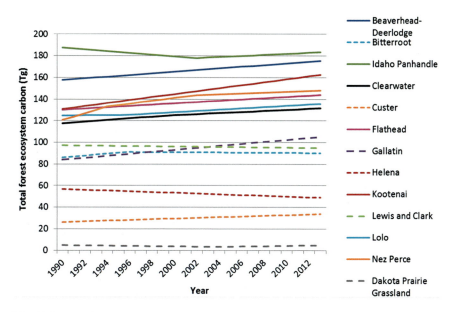

Fig. 10.2 Total forest ecosystem carbon for national forests and grasslands in the U.S. Forest Service Northern Region, 1990–2013

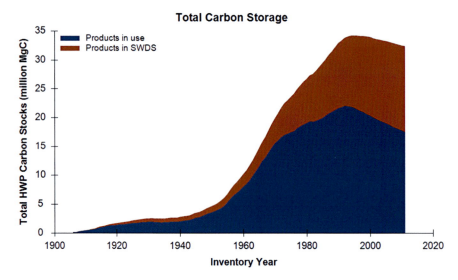

Fig. 10.3 Cumulative total carbon stored in harvested wood products (HWP) manufactured from national forests in the U.S. Forest Service Northern Region, including carbon in products still in use and stored at solid waste disposal sites (SWDS) (Stockmann et al. 2014)

harvest. The largest impact on carbon storage in Idaho Panhandle National Forest was disease, followed by harvest.

Elevated nitrogen deposition, a phenomenon observed across much of the western United States, may increase wood production and accumulation of soil organic matter, thus increasing carbon sequestration. Carbon uptake in living biomass is often a transitory phenomenon, but carbon accumulation in soil is potentially a long-term sink because belowground carbon has longer turnover times than aboveground carbon (Bytnerowicz et al. 2007).

Fungal pathogens, especially various types of root rot, are another key concern for forests and may affect the ability of forests to sequester carbon (Hicke et al. 2012). Increased temperature and humidity coupled with decreased snow and cold weather facilitate the spread of some root rots. As more trees die and decompose, forests could switch from carbon sinks to carbon sources.

Adaptive capacity for sequestering carbon depends on the spatial and temporal scales at which this ecosystem service is defined. Carbon storage in any particular forest location may go up or down over time, but analysis of storage should logically occur at very large spatial scales. Adaptive capacity for carbon storage is probably low, because most of the factors affecting carbon sequestration are external, especially wildfire and other disturbances that can override other factors, including management.

10.4.9 Summary

Ecosystem services are the benefits people derive from landscapes and encompass the values that motivate people to live in the Northern Rockies. Ecosystem services are the core of our sense of place. Social (demographic changes, economics, policy) and environmental (e.g., climate change) factors, individually and interactively, can affect ecosystem services both positively and negatively. Opportunities for adaptation to climate change need to consider the broader social context to be successful and to set priorities. In summary:

- **Water yield** is not expected to change significantly, and few communities are likely to experience water shortages and water stress. The biggest effect on water yield will be timing of water availability, although this could potentially be overcome with reservoir storage. Because agriculture is the largest consumer of water and a big economic force, rural agricultural communities will be disproportionately affected by climate change.
- **Water quality** is closely tied to water yield. Increased occurrence of wildfires and floods will add sediment to rivers and reservoirs, affecting instream water quality and making water treatment more expensive. Agriculture is the major source of impairment, leading to degraded riparian and aquatic habitat, increased water temperatures, and high levels of contamination. Climate change is expected to amplify these effects.
- **Wood products** provide jobs in the region. Climate change will cause more wildfires and insect outbreaks, but effects on wood products will be smaller than from economic forces and policies. Timber production has been in steady decline, a trend that will probably continue, with significant effects on economic vitality of small rural towns. Diversification of local economies can help buffer the loss of timber-related jobs.
- **Minerals and mineral extraction** are important economic drivers, and are not vulnerable to climate change. However, mineral and energy extraction are connected to other ecosystem services, particularly water quality. Wildfires, floods, and mudslides all put mineral extraction infrastructure and associated watersheds at risk.
- **Forage for livestock** may benefit from increased productivity in a warmer climate, with minor economic benefits to ranching in small communities. Most stressors on grazing are human induced, including loss of rural population, spread of nonnative plant species, and fragmentation of rangelands.
- **Viewsheds** and **air quality** will be affected by increasing wildfire frequency and length of pollen seasons. A growing percentage of the region's population will be in at-risk demographic groups who will suffer respiratory and other medical problems on days with poor air quality.
- The ability to **regulate soil erosion** will be diminished by agricultural expansion, spread of invasive plants, and increased frequency of wildfire and floods. Increased capital investments may be needed for water treatment plants if water

quality degrades significantly. Best practices in agriculture and road construction can mitigate some effects.
- **Carbon sequestration** will be challenged by increasing wildfires and insect outbreaks, especially in the western part of the Northern Rockies. Managing forests for carbon sequestration is likely to become more important in response to national climate policies, but will need to be implemented in the context of other resource objectives. Thinning and fuel treatments can help reduce the magnitude of periodic carbon pulses.

References

American Lung Association. (2015). *State of the air 2015*. http://www.stateoftheair.org. 28 May 2015.

Bedsworth, L. (2011). Air quality planning in California's changing climate. *Climatic Change, 111*, 101–118.

Beyers, W. B., & Nelson, P. B. (2000). Contemporary development forces in the nonmetropolitan west: New insights from rapidly growing communities. *Journal of Rural Studies, 16*, 459–474.

Borrie, W. T., Davenport, M., Freimund, W. A., & Manning, R. E. (2002). Assessing the relationship between desired experiences and support for management actions at Yellowstone National Park using multiple methods. *Journal of Park and Recreation Administration, 20*, 51–64.

Brandt, J. P., Morgan, T. A., Keegan, C. E., et al. (2012). *Idaho's forest products industry and timber harvest, 2006, Resource Bulletin RMRS-RB-12*. Fort Collins: U.S. Forest Service, Rocky Mountain Research Station.

Bytnerowicz, A., Omasa, K., & Paoletti, E. (2007). Integrated effects of air pollution and climate change on forests: A northern hemisphere perspective. *Environmental Pollution, 147*, 438–445.

Center for Disease Control. (2009). *BRFSS [Behavioral Risk Factor Surveillance System] annual survey data*. Atlanta: U.S. Department of Health and Human Services, Center for Disease Control. http://www.cdc.gov/brfss. 28 May 2015.

Cleland, E. E., Collins, S. L., Dickson, T. L., Farrer, E.C., Gross, K. L., Gherardi, L. A., Hallett, L. M., et al. (2013). Sensitivity of grassland plant community composition to spatial vs. temporal variation in precipitation. Ecology, 94(8), 1687–1696.

Collins, S., & Larry, E. (2007). Caring for our natural assets: An ecosystem services perspective. In R. L. Deal (Ed.), *Integrated restoration of forested ecosystems to achieve multiresource benefits: Proceedings of the 2007 national silviculture workshop, General Technical Report PNW-733* (pp. 1–11). Portland: U.S. Forest Service, Pacific Northwest Research Station.

Cutter, S. L., Boruff, B. J., & Shirley, W. L. (2003). Social vulnerability to environmental hazards. *Social Science Quarterly, 84*, 242–261.

Fothergill, A., & Peek, L. A. (2004). Poverty and disasters in the United States: A review of recent sociological findings. *Natural Hazards, 32*, 89–110.

Foti, R., Ramirez, J. A., & Brown, T. C. (2012). *Vulnerability of U.S. water supply to shortage: A technical document supporting the Forest Service 2010 RPA assessment* (General Technical Report RMRS-GTR-295). Fort Collins: U.S. Forest Service, Rocky Mountain Research Station.

Healey, S. P., Urbanski, S. P., Patterson, P. L., et al. (2014). A framework for simulating map error in ecosystem models. *Remote Sensing of the Environment, 150*, 207–217.

Hicke, J. A., Allen, C. D., Desai, A. R., et al. (2012). Effects of biotic disturbances on forest carbon cycling in the United States and Canada. *Global Change Biology, 18*, 7–34.

Irland, L. C., Adams, D., Alig, R., et al. (2001). Assessing socioeconomic impacts of climate change on US forests, wood-product markets, and forest recreation. *Bioscience, 51*, 753–764.

Johnson, W. C., Millett, B. V., Gilmanov, T., et al. (2005). Vulnerability of northern prairie wetlands to climate change. *Bioscience, 55*, 863–872.

Karetinkov, D., Parra, N., Bell, B., Ruth, M., Ross, K., & Irani, D. (2008). Economic impacts of climate change on North Dakota. A review and assessment conducted by the Center for Integrative Environmental Research (CIER).

Keegan, C. E., Sorenson, C. B., Morgan, T. A., et al. (2012). Impact of the Great Recession and housing collapse on the forest products industry in the western United States. *Forest Products Journal, 61*, 625–634.

Keegan, C. E., Todd, A. M., Wagner, F. G., et al. (2005). Capacity for utilization of USDA Forest Service, Region 1 small-diameter timber. *Forest Products Journal, 55*, 143–147.

Kinney, P. L. (2008). Climate change, air quality, and human health. *American Journal of Preventive Medicine, 35*, 459–467.

Kline, J. D., & Mazzotta, M. J. (2012). *Evaluating tradeoffs among ecosystem services in the management of public lands* (General Technical Report PNW-GTR-865). Portland: U.S. Forest Service, Pacific Northwest Research Station.

Kunce, M., & Shogren, J. F. (2005). On interjurisdictional competition and environmental federalism. *Journal of Environmental Economics and Management, 50*, 212–224.

Maher, A. T. (2007). *The economic impacts of sagebrush steppe wildfires on an eastern Oregon ranch.* Master's thesis. Corvallis: Oregon State University.

McIver, C. P., Sorenson, C. B., Keegan, C. E., et al. (2013). *Montana's forest products industry and timber harvest 2009* (Resource Bulleting RMRS-RB-16). Fort Collins: U.S. Forest Service, Rocky Mountain Research Station.

Millennium Ecosystem Assessment. (2005). *Ecosystems and human well-being: Synthesis.* Washington, DC: Island Press.

Montana Department of Natural Resources and Conservation (Montana DNRC). (2014). Montana state water plan draft August 21, 2014. Helena. http://leg.mt.gov/content/Committees/Interim/2013-2014/EQC/Meetings/September-2014/state-water-plan-draft-lowres.pdf. 28 May 2015.

Murdoch, P. S., Baron, J. S., & Miller, T. L. (2000). Potential effects of climate change on surface-water quality in North America. *Journal of the American Water Resources Association, 36*, 347–366.

North Dakota Wheat Commission. (2007). *Report to the 2007 North Dakota legislative assembly: Economic importance of wheat.* http://www.ndwheat.com/uploads%5Cresources%5C614%5C071legreport.pdf. 28 May 2015.

Pan, Y., Birdsey, R., Hom, J., & McCullough, K. (2009). Separating effects of changes in atmospheric composition, climate and land-use on carbon sequestration of U.S. mid-Atlantic temperate forests. *Forest Ecology and Management, 259*, 151–164.

Pan, Y., Birdsey, R. A., Fang, J., et al. (2011). A large and persistent carbon sink in the world's forests. *Science, 333*, 988–993.

Polley, H. W., Jin, V. L., & Fay, P. A. (2012). Feedback from plant species change amplifies CO_2 enhancement of grassland productivity. *Global Change Biology, 18*, 2813–2823.

Power, T. M. (1998). *Lost landscapes and failed economies: The search for a value of place.* Washington, DC: Island Press.

Rasker, R. (1993). A new look at old vistas: The economic role of environmental quality in Western public lands. *University of Colorado Law Review, 65*, 369–399.

Raymond, C. L., Healey, S. P., Peduzzi, A., & Patterson, P. L. (2015). Representative regional models of post-disturbance forest carbon accumulation: Integrating inventory data and a growth and yield model. *Forest Ecology and Management, 336*, 21–34.

Reeves, M. C., & Mitchell, J. E. (2012). *A synoptic review of U.S. rangelands: A technical document supporting the Forest Service 2010 RPA assessment* (General Technical Report RMRS-GTR-288). Fort Collins: U.S. Forest Service, Rocky Mountain Research Station.

Riebsame, W. E., Hannah, G., Theobald, D., et al. (1997). *Atlas of the New West: Portrait of a changing region.* New York: Norton.

Rosenzweig, C., Iglesias, A., Yang, X. B., et al. (2000). *Climate change and U.S. agriculture: The impacts of warming and extreme weather events on productivity, plant diseases, and pests.* Cambridge, MA: Harvard Medical School.

Ryan, M. G., Archer, S. R., Birdsey, R., et al. (2008). Land resources. In M. Walsh, P. Backlund, A. Janetos, & D. Schimel (Eds.), *The effects of climate change on agriculture, land resources, water resources, and biodiversity in the United States* (pp. 75–120). Washington, DC: Climate Change Science Program, Subcommittee on Global Change Research.

Schindler, D. W., Hecky, R. E., Findlay, D. L., et al. (2008). Eutrophication of lakes cannot be controlled by reducing nitrogen input: Results of a 37-year whole-ecosystem experiment. *Proceedings of the National Academy of Sciences, USA, 105*, 11254–11258.

Sham, C. H., Tuccillo, M. E., & Rooke, J. (2013). *Effects of wildfire on drinking water utilities and best practices for wildfire risk reduction and mitigation, Web Report 4482.* Denver: Water Research Foundation.

Skog, K. E. (2008). Sequestration of carbon in harvested wood products for the United States. *Forest Products Journal, 58*, 56–72.

Sorenson, L. G., Goldberg, R., Root, T. L., & Anderson, M. G. (1998). Potential effects of global warming on waterfowl populations breeding in the northern Great Plains. *Climatic Change, 40*, 343–369.

Stockmann, K., Anderson, N., Young, J., et al. (2014). *Estimates of carbon stored in harvested wood products from the U.S. Forest Service Northern Region, 1906–2012* (Unpublished report). Missoula: U.S. Forest Service, Rocky Mountain Research Station.

Thamke, J. N., & Smith, B. D. (2014). *Delineation of brine contamination in and near the East Poplar oil field, Fort Peck Indian Reservation, northeastern Montana, 2004–09, U.S. Geological Survey Scientific Investigations Report 2014–5024.* Reston: U.S. Geological Survey.

U.S. Forest Service (USFS). (2009). *Watershed and air management, Forest Service Manual 2550, Amendment 2500-2009-1.* Washington, DC: U.S. Forest Service.

U.S. Forest Service (USFS). (2015). *Baseline estimates of carbon stocks in forests and harvested wood products for National Forest System units, Northern Region* (unpublished report). http://www.fs.fed.us/climatechange/documents/NorthernRegionCarbonAssessment.pdf. 1 Apr 2016.

Van Vliet, M. T. H., Ludwig, F., Zwolsman, J. J. G., et al. (2011). Global river temperatures and sensitivity to atmospheric warming and changes in river flow. *Water Resources Research, 47*, W02544.

Wagner, T., Congleton, J. L., & Marsh, D. M. (2004). Smolt-to-adult return rates of juvenile Chinook salmon transported through the Snake-Columbia River hydropower system, USA, in relation to densities of co-transported juvenile steelhead. *Fisheries Research, 68*, 259–270.

Warziniack, T. (2014). A general equilibrium model of ecosystem services in a river basin. *Journal of the American Water Resources Association, 50*, 683–695.

Woodall, C., Smith, J., & Nichols, M. (2013). *Data sources and estimation/modeling procedures for the National Forest System carbon stock and stock change estimates derived from the U.S. National Greenhouse Gas Inventory.* http://www.fs.fed.us/climatechange/documents/NFSCarbonMethodology.pdf. 1 Apr 2016.

Chapter 11
Effects of Climate Change on Cultural Resources in the Northern Rockies

Carl M. Davis

Abstract Cultural resources in the Northern Rockies are currently vulnerable to various natural and human agencies, including wildfire and biological processes, vandalism and other depreciative human behaviors, and changing population demographics and recreational use. Climate change has the potential to accelerate some of these ongoing effects to cultural resources. Increasing wildfires will have a direct effect on cultural resources, because they are broadly distributed throughout forest and grassland ecosystems. Melting ice caused by climate change poses a risk to previously ice-encased and well-preserved cultural resources. Seasonal aridity and prolonged drought will accelerate soil deflation and erosion, and expose archaeological sites once buried in prairie or mountain soils. At the same time, a projected increase in winter precipitation, coupled with earlier and more intense spring runoff, poses another threat to cultural resources. Climate-induced changes in terrestrial and aquatic habitats also affect abundance of culturally-valued plants, animals and fish, affecting the ability of Native American tribes to exercise their treaty rights.

Damage to cultural and historic sites is irreversible, making protection a key management focus. To some extent, wildfire effects can be mitigated through active prevention measures (for example, thinning trees around historic structures) and fire suppression and recovery tactics. Hydrological events are unpredictable, and protection measures such as stabilization and armoring are expensive. Nonetheless, federal agencies have a strong mandate to implement measures to protect cultural sites threatened by such natural processes and emergency events. Survey and evaluation in areas where cultural resources are concentrated or likely is ongoing, although intermittent, in the Northern Rockies. It will be possible to locate and monitor cultural resources only if these efforts are significantly expanded.

Keywords American Indians • Artifacts • Cultural landscapes • Cultural resources • Historic buildings • National Historic Preservation Act • Traditional cultural uses

---

C.M. Davis (✉)
U.S. Forest Service, Northern Region, Missoula, Montana, USA
e-mail: cmdavis1134@icloud.com

11.1 Background and Cultural Context

People have inhabited the Northern Rocky Mountains and Great Plains of the United States since the end of the last Pleistocene glacial period (Fagan 1990; Meltzer 2009), and evidence of this distant and more recent human occupation is found throughout the assessment area. The Northern Rockies and Plains are the ancestral homeland, aboriginal territory, and hunting ground of the Assiniboine, Blackfeet, Chippewa-Cree, Crow, Hidatsa, Kiowa, Kutenai, Nez Perce, Northern Cheyenne, Salish, Shoshone, Sioux and other Plains, Intermountain, and Columbia Plateau American Indian tribes (Walker 1988; Schleiser 1994; DeMallie 2001). Beginning in the eighteenth century, the region was explored and then settled by people of many different European ancestries (White 1993). The region has always contained a diversity of cultural backgrounds and lifeways.

Archaeological and historical evidence of past cultural groups, interactions, and events are termed "cultural resources," and include (1) ancient Indian camps and villages, rock art, tool stone quarries, and travel routes, (2) historic military forts and battlefields, mining and logging ruins, and homesteads, and (3) ranger stations, fire lookouts, and recreation sites built by the Civilian Conservation Corps (NPS 2015b). The U.S. Forest Service (USFS) Northern Region alone has documented approximately 20,000 cultural resources, representing a small fraction of what likely exists across the entire assessment area.

Protection of cultural resources has been formally recognized since 1906 when the Antiquities Act was signed into law, and has been reaffirmed by the Historic Sites Act of 1935, the National Historic Preservation Act of 1966, the Archaeological Resources Protection Act of 1979 and the Native American Graves Protection and Repatriation Act of 1990. Federal land management agencies are required to identify, evaluate and preserve historic, scientific, commemorative, and cultural values of archaeological and historic sites and structures on public lands for present and future generations (NPS 2015a; USFS 2008). The President of the United States has authority to designate national monuments in order to protect landmarks, structures, and objects of historical or scientific significance. In 1966, Congress declared it to the our national policy that the Federal government will "administer federally owned, administered, or controlled prehistoric and historic resources in a spirit of stewardship for the inspiration and benefit of present and future generations." Thus, a core mission of the National Park Service is the preservation, enhancement and interpretation of cultural resources. The USFS and other federal land management agencies protect and manage cultural resources as part of their multiple use missions.

Protection of cultural resources also includes ongoing use of resources and associated activities relevant to the continuation of extant American Indian and other cultures (NPS 2011). Many cultural resources are currently vulnerable to natural biophysical phenomena and human activities. Wildfire and decomposition degrade and destroy cultural resources, particularly those made of wood or located in erosion-prone environments. Vandalism, illegal artifact collecting, arson, and other human behaviors also damage cultural resources. Land management actions can affect cultural sites, although federal land managers attempt to protect and mitigate adverse effects wherever possible.

11.2 Climate Change Effects on Cultural Resources

11.2.1 Primary Effects and Stressors

This evaluation of the potential effects of climate change on cultural resources in the Northern Rockies is quite general, because so little information has been generated on this topic compared to the effects of climate change on natural resources, and because it is difficult to infer the spatial extent and timing of specific effects. Inferences in this chapter are based on a synthesis of relevant literature from different disciplines to project how an altered climate, both directly and indirectly (through increased disturbance), will create conditions that modify the condition of and access to cultural resource sites.

Climate change has the potential to exacerbate and accelerate existing effects on cultural resources (Rockman 2015; Morgan et al. 2016) (Table 11.1). A warmer climate will alter the scale of wildfires across western North America (McKenzie et al. 2004; Schoennagel et al. 2004; Chap. 7), thus having at least three general effects on cultural resources. First, wildfires burn cultural resources made of wood and

Table 11.1 Summary of climate change stressors and potential effects on cultural resources in the Northern Rockies

Climate change stressor	Biophysical effects	Effects on cultural sites and landscapes
Temperature increase	Wildfire	Combustion, damage, destruction
	Drought, erosion	Exposed artifacts and cultural features
	Vegetation changes	Altered physical appearance, integrity
	Spread of invasive species	Altered physical appearance, integrity
	Ice patch melt	Artifact decay and theft
	Altered freeze-thaw cycles	Saturation, desiccation, warping, biochemical changes
Altered precipitation	Earlier seasonal runoff, flooding	Removal, damage, degradation
	Debris flows, slumping	Burial, removal, degradation
	Down cutting, mass wasting	Removal, damage, degradation
	Increased moisture and humidity	Decay, oxidation, exfoliation, corrosion, biochemical changes
	Extreme precipitation events	Removal, damage, degradation, collapse, exposure

For additional detail, see UNESCO (2007), Rockman (2015), and Morgan et al. (2016)

Fig. 11.1 Aboriginal stone cairn exposed by wildfire in Custer-Gallatin National Forest (Photo by Halcyon LaPoint, U.S. Forest Service)

combustible materials, including ancient wood shelters and game drives, and historic homesteads, mining ruins, and early USFS backcountry cabins, lookouts and administrative structures. Second, wildfire suppression and post-fire recovery operations (e.g., heavy equipment use, erosion abatement) affect standing structures and archaeological sites buried in forest soils. Third, flooding and debris flows can damage (e.g., erode away, disturb, bury) cultural resources exposed in the post-fire environment. However, fire can be beneficial if it exposes cultural sites that were not previously visible and archaeologists have the opportunity to record them and develop protection measures where they are now threatened by natural or human disturbances (Fig. 11.1).

Federal agencies implement various actions to reduce the effects of wildfire on cultural resources, such as encasing historic structures in fireproof wrap, reducing suppression activities near cultural sites, and physical armoring of cultural resources vulnerable to post-fire flooding. Because it is difficult and expensive to implement these actions across large landscapes, damage is expected to continue as climate change amplifies area burned.

Seasonal aridity and prolonged drought accelerate soil deflation and erosion, exposing archaeological sites buried in the soil. Wind and water erosion can remove ground cover, revealing artifacts and features such as cooking hearths and tool-making areas. Newly exposed artifacts make them vulnerable to illegal collecting, which can be intensified in areas where livestock grazing, recreation, and mining have already caused impacts. For example, livestock often converge around streams and natural springs where archaeological sites of ancient hunter-gatherers are common.

11 Effects of Climate Change on Cultural Resources in the Northern Rockies

Fig. 11.2 Post-wildfire debris flow that obliterated or covered cultural resources in Meriwether Canyon, Helena National Forest (Photo by Carl Davis, U.S. Forest Service)

Periods of dry climate and drought have occurred throughout the Holocene in the intermountain West, with corresponding episodes of soil deflation, erosion, and down cutting of drainages and stream beds (Meltzer 1990; Ruddiman 2007). Warmer temperatures in the future (Mayewski and White 2002; IPCC 2007; Chap. 2) will create additional potential for cultural resource loss through drought and erosion, particularly in drier areas such as southeastern Montana.

If winter precipitation increases (Chap. 2) and reduced snowpack leads to higher winter streamflows (Chap. 3), archaeological and historic sites will be increasingly vulnerable to flooding, debris flows, down cutting, and mass wasting of underlying landforms. This already occurs in the aftermath of large wildfires, especially in the dry mountain ranges of central and eastern Montana (Fig. 11.2), and an increase in extreme events (Chap. 7) will almost certainly increase hydrological impacts on cultural resources (National Research Council 2002).

Persistent high-elevation snowfields contain artifacts remaining from hunting and gathering forays by Native Americans in mountain environments hundreds to thousands of years ago (Lee 2012) (Fig. 11.3). Melting ice caused by a warmer climate poses a risk to previously ice-encased cultural resources that are well preserved. For example, ancient bone, wood, and fiber artifacts have been revealed by melting ice patches in the Beartooth Mountains (south-central Montana). Melting ice provides opportunities for archaeologists and Native Americans to locate, document, and archive artifacts, but it also makes artifacts susceptible to decay or theft.

Fig. 11.3 Melting perennial ice patches expose prehistoric artifacts in Custer-Gallatin National Forest (Photo by Craig Lee, Montana State University)

Climate change can affect cultural landscapes whose integrity is derived from both cultural resources and environmental context (NPS 1994; Rockman 2015), including ancient American Indian travel routes, battlefields (e.g., Big Hole Battlefield) and historic mining districts. Altered distribution and abundance of dominant vegetation could potentially affect the physical and visual integrity of such landscapes (Melnick 2009). For example, whitebark pine (*Pinus albicaulis*) is an important historical component of the Alice Creek-Lewis and Clark Pass (Road to the Buffalo Trail) cultural landscape on the Continental Divide near Helena, Montana (Fig. 11.4). Whitebark pine is currently in decline, because warmer temperatures have facilitated extensive outbreaks of mountain pine beetle (*Dendroctonus ponderosae*) in addition to several decades of mortality and damage from white pine blister rust (*Cronartium ribicola*), a nonnative fungal pathogen (Tomback and Kendall 2001) (Chap. 7).

Cultural sites and landscapes are also recognized for their traditional importance to descendant communities, particularly Native American tribes, who value traditional-use areas for foods, medicinal and sacred plants, paints, and ceremonial and spiritual places. Significant climate-induced effects in these landscapes, particularly shifts in native vegetation, may reduce and even sever the continuous cultural connectivity of these areas by indigenous peoples and local communities.

Extreme events related to climate change (e.g., wildfire, flooding, debris flows) may affect historic buildings and structures. In addition to these direct effects, period furniture, interpretive media, and artifact collections inside historic (and

Fig. 11.4 Whitebark pine mortality may affect the integrity and status of cultural sites, such as the Lewis and Clark Pass cultural landscape and National Register District shown here. Significant landscape change may also affect indigenous peoples and local communities who use the area and its resources (Photo by Sara Scott, Montana Department of Fish, Wildlife, and Parks)

non-historic) buildings can be affected by extreme events. Additional stressors include increased heat, moisture, humidity, freeze-thaw events, insect infestation, and fungi, all of which can accelerate weathering, deterioration, corrosion, and decay of structures and ruins (UNESCO 2007).

Climate change may also diminish the appeal of cultural sites and cultural landscapes for members of the public who visit these sites and interpretive exhibits. Extensive outbreaks of mountain pine beetle and other insects, which have been facilitated by higher temperature, have turned some historic landscapes in southwestern Montana from green to brown to gray (e.g., Logan and Powell 2001). Dead and dying forests also present hazards to hikers, sightseers, and other visitors (Chap. 9). Altered ecological conditions surrounding cultural sites and within cultural landscapes may reduce their attractiveness and value for tourism, recreation and other purposes, with potential impact on local communities and economies (Chaps. 9 and 10).

11.2.2 Spatial and Temporal Risk Assessment

Climate change effects on cultural resources will be highly variable across the assessment area, with some effects occurring within the next few decades and others by the end of the twenty-first century. Wildfire will be the biggest and most pervasive risk for cultural resources on federal lands in the region, creating a mosaic of burned areas of different sizes and severities over time. For example, large, high-severity wildfires since 2000 have burned hundreds of thousands of hectares on national forests in Idaho and Montana, from the Bitterroot National Forest in western Montana to the Custer Gallatin National Forest in southeastern Montana (Fig. 11.1). Hundreds of cultural resources have been affected. Glacier and Yellowstone National Parks, Bureau of Land Management units, and other public lands have also experienced large fires since the 1990s, with the same consequences to cultural resources. This is likely to continue into the foreseeable future.

Increased aridity and drought may be partly offset if winter precipitation increases in the future (Chap. 2), making it difficult to quantify the long-term effects of extreme weather and hydrologic events. Resource damage will be greatest in areas prone to hydrologic disturbance, such as canyon mouths and river bottoms where cultural sites are often concentrated. It will be difficult to armor and protect sites in these locations, and artifact collectors may target these areas where cultural materials are exposed in newly disturbed landforms or are deposited there by water and debris.

Other effects on cultural resources will be more subtle or slow to appear. Shifting vegetation distribution and abundance will occur gradually over many decades, typically requiring one or more large disturbances to promote regeneration. Climate change effects to historic buildings and structures will be gradual and cumulative (deterioration, decay) in some cases, and sudden and direct (e.g., structural collapse caused by snow loading and excessive moisture) in others (Morgan et al. 2016). Some natural resources associated with traditional cultural landscapes, still used by Native Americans today, may deteriorate or disappear. However, increased wildfire may increase the abundance of some culturally valuable species, such as huckleberry (*Vaccinium* spp.), common camas (*Camassia quamash*), and nodding onion (*Allium cernuum*).

The effects of climate change on cultural resource tourism are difficult to estimate because tourism is strongly affected by many social and economic factors, but it is unlikely that most popular cultural sites will completely deteriorate in the next several decades. Visiting historical sites is popular throughout the intermountain West (Nickerson 2014), and tourism is an important economic contributor to many local communities (Chap. 9). The direct effects of hot weather could reduce public interest in visiting cultural landscapes and interpretive sites, particularly in areas recently affected by dying and dead vegetation, severe wildfires, or floods, with secondary economic effects on local communities.

11.3 Adapting Cultural Resources and Management to Climate Change

Federal agencies have the capacity to address some of the projected effects of climate change on cultural resources. Fuels reduction around significant cultural resources is already in place in some locations, thus reducing the intensity and

Fig. 11.5 The Bar Gulch Cabin (Helena National Forest) survived a wildfire in 2000, because it was protected by fire retardant wrap and a water sprinkler system. Historic structures will be vulnerable if fire frequency increases in the future (Photo by Carl Davis, U.S. Forest Service)

severity of future wildfires. Heritage personnel in national forests and national parks are engaged in all aspects of wildfire management, helping to protect cultural resources that could be damaged by wildfires, fire suppression and fire recovery (Fig. 11.5). Fire vulnerability assessment and abatement programs for cultural resources deserves additional emphasis in anticipation of more wildfires in the future.

Less progress has been made in implementing protection strategies for cultural resources in areas prone to large-scale hydrologic events, partly because the scope of this risk has not been documented. Flooding and geomorphic disturbance are unpredictable in space and time. Protection measures (e.g., stabilization, armoring) are often prohibitively expensive, typically requiring expertise in hydrology, engineering, and other disciplines to develop effective solutions.

Survey, inventory, and evaluation in areas where cultural resources are concentrated or likely to exist are ongoing, albeit at a low level of activity. Identification and monitoring of at-risk resources will be possible only if these efforts are greatly expanded. High-elevation melting ice patches are currently a priority, but surveys are also needed in locations where artifacts may be damaged by water and earth movement (e.g., canyon and foothills areas). Areas with cultural resources can be correlated with areas where flooding and ice melt are expected will help identify landscapes at greatest risk.

Potential climate-induced vegetation shifts in cultural landscapes could be partly mitigated through silvicultural treatments and prescribed burning, although the effectiveness of proposed treatments relative to the scope and scale of the cultural landscape is difficult to evaluate. Careful monitoring and tracking of vegetation stability and change in cultural landscapes will become increasingly important in future decades.

The potential effects of climate change on the historic built environment in the Northern Rockies has received little attention. However, some actions may eventually be necessary to reduce the potentially negative effects of climate change on historic buildings and structures. These actions could include hazardous fuels reduction, flood and erosion control, insect abatement, building weatherization, and structural stabilization. Conducting vulnerability assessments is the first step to planning any remediation work. Collaborative efforts that include agency managers, heritage specialists, historic building preservation teams, partners, and volunteers are needed to develop priorities and initiate this work on public lands.

References

DeMallie, R. J. (Ed.). (2001). *Handbook of North American Indians: Plains, volume 13*. Washington, DC: Smithsonian Institution.

Fagan, B. M. (1990). *The journey from Eden: The peopling of our world*. London: Thames & Hudson.

Intergovernmental Panel on Climate Change (IPCC). (2007). The physical science basis. Contribution of Working Group I to the Fourth Assessment Report of the Intergovernmental Panel on Climate Change. S. Solomon, D. Qin, M. Manning, et al. (Eds.) Cambridge, UK: Cambridge University Press.

Lee, C. M. (2012). Withering snow and ice in the mid-latitudes: A new archaeological and paleobiological record for the Rocky Mountain region. *Arctic, 65*, 165–177.

Logan, J., & Powell, J. (2001). Ghost forests, global warming, and the mountain pine beetle (Coleoptera: Scolytidae). *American Entomologist, 47*(3), 160–173.

Mayewski, P. A., & White, F. (2002). *The ice chronicles: The quest to understand global climate change*. Hanover, NH: University of New Hampshire Press.

McKenzie, D., Gedalof, Z., Peterson, D. L., & Mote, P. (2004). Climatic change, wildfire, and conservation. *Conservation Biology, 18*(4), 890–902.

Melnick, R. Z. (2009). Climate change and landscape preservation: A twenty-first century conundrum. *APT Bulletin: Journal of Preservation Technology, 40*(3-4), 34–43.

Meltzer, D. J. (1990). Human responses to middle Holocene (Altithermal) climates on the North American Great Plains. *Quaternary Research, 52*, 404–416.

Meltzer, D. J. (2009). *First peoples in a new world: Colonizing ice age America*. Berkeley: University of California Press.

Morgan, M., Rockman, M., Smith, C., & Meadow, A. (2016). *Climate change impacts on cultural resources* (Cultural resources partnerships and science). Washington, DC: National Park Service.

National Park Service (NPS). (1994). *Protecting cultural landscapes: Planning, treatment and management of historic landscapes* (Preservation Brief 36). Washington, DC: National Park Service.

National Park Service (NPS). (2011). Cultural resources, partnerships and science directorate. http://www.nps.gov/history/tribes/aboutus.htm. 30 Jan 2017.

National Park Service (NPS). (2015a). Archaeology program–Antiquities Act 1906–2006. http://www.nps.gov/archeology/sites/antiquities/about. 30 Jan 2017.

National Park Service (NPS). (2015b.) Glacier National Park: What are cultural resources? http://gnpculturalresourceguide.info/files/resources/What%20Are%20Cultural%20ResourcesFinal.pdf. 30 Jan 2017.

National Research Council. (2002). *Abrupt climate change: Inevitable surprises*. Washington, DC: National Academy Press.

Nickerson, N. P. (2014). *Travel and recreation in Montana: 2013 review and 2014 outlook*. Missoula: University of Montana, College of Forestry and Conservation.

Rockman, M. (2015). An NPS framework for addressing climate change with cultural resources. *The George Wright Forum, 32*(1), 37–50.

Ruddiman, W. F. (2007). *Earth's climate: Past and future*. New York: W. H. Freeman.

Schleiser, K. H. (1994). *Plains Indians, A.D. 500-1500: The archaeological past of historic groups*. Norman: University of Oklahoma Press.

Schoennagel, T., Verblen, T. T., & Romme, W. H. (2004). The interaction of fire, fuels, and climate across Rocky Mountain forests. *Bioscience, 54*(7), 661–676.

Tomback, D. F., & Kendall, K. C. (2001). Biodiversity losses: The downward spiral. In D. F. Tomback, S. F. Arno, & R. E. Keane (Eds.), *Whitebark pine communities: Ecology and restoration* (pp. 243–262). Washington, DC: Island Press.

U.S. Forest Service (USFS). (2008). *Forest service manual, FSM-recreation, wilderness, and related resource management, chapter 2360-heritage program management*. Washington, DC: National Headquarters.

United Nations Educational, Scientific, and Cultural Organization (UNESCO). (2007). *Climate change and world heritage: Report on predicting and managing the impacts of climate change on world heritage and strategy to assist states parties to implement appropriate management responses* (World Heritage Report 22). Paris: UNESCO, World Heritage Centre.

Walker, D. E. (Ed.). (1988). *Handbook of North American Indians: Plains, volume 12*. Washington, DC: Smithsonian Institution.

White, R. (1993). *It's your misfortune and none of my own: A new history of the American West*. Norman: University of Oklahoma Press.

Chapter 12
Toward Climate-Smart Resource Management in the Northern Rockies

Jessica E. Halofsky, David L. Peterson, S. Karen Dante-Wood, and Linh Hoang

Abstract The Northern Rockies Adaptation Partnership facilitated the largest climate change adaptation effort on public lands to date, including participants from federal agencies and stakeholder organizations interested in a broad range of resource issues. It achieved specific goals of national climate change strategies for the U.S. Forest Service and National Park Service, providing a scientific foundation for resource management and planning in the Northern Rockies. The large number of adaptation strategies and tactics, many of which are a component of current management practice, provide a pathway for slowing the rate of deleterious change in resource conditions. Rapid implementation of adaptation—in land management plans, National Environmental Policy Act documents, project plans, and restoration—will help maintain functionality of terrestrial and aquatic ecosystems in the Northern Rockies, as well as build the organizational capacity of federal agencies to incorporate climate change in their mission of sustainable resource management. Long-term monitoring will help detect potential climate change effects on natural resources, and evaluate the effectiveness of adaptation options that have been implemented.

Keywords Adaptation • Management planning • Implementation • Science-management partnerships • Organizational capacity

---

J.E. Halofsky (✉)
School of Environmental and Forest Sciences, University of Washington, Seattle, WA, USA
e-mail: jhalo@uw.edu

D.L. Peterson
U.S. Forest Service, Pacific Northwest Research Station, Seattle, WA, USA
e-mail: peterson@fs.fed.us

S. Karen Dante-Wood
U.S. Forest Service, Office of Sustainability and Climate, Washington, DC, USA
e-mail: skdante@fs.fed.us

L. Hoang
U.S. Forest Service, Northern Region, Missoula, MT, USA
e-mail: lhoang@fs.fed.us

12.1 Partnership and Process

The Northern Rockies Adaptation Partnership (NRAP) provided the scientific documentation for understanding and responding to climate change in Northern Rockies ecosystems. The assessment team synthesized scientific information to evaluate climate change vulnerability, working with resource managers to (1) develop adaptation options that reduce the negative effects of climate change, and (2) transition ecosystems and organizations to a permanently warmer world. Vulnerability assessment information and adaptation options developed by the NRAP are now being used to support national forests and national parks in implementing agency climate change strategies (NPS 2010; USFS 2010, 2012; Chap. 1). In addition, we catalyzed a collaboration of 35 land management agencies and stakeholders with common interests in addressing climate change in the Northern Rockies, an enduring partnership that will ensure timely and consistent application of the assessment in the years ahead.

12.1.1 Increasing Organizational Capacity to Address Climate Change

Although the NRAP was led primarily by federal agencies, the assessment information that was developed should be relevant for other land management agencies, tribes, and nongovernmental organizations in the Northern Rockies. This process can be replicated and implemented by other organizations, and the adaptation options can be used in the Northern Rockies and beyond. Like previous adaptation projects (Halofsky et al. 2011; Swanston et al. 2011; Raymond et al. 2014; Halofsky and Peterson 2016, 2017), a science-management partnership was critical to the success of the Northern Rockies effort. For others interested in emulating this approach, we encourage them to pursue this type of partnership as the foundation for increasing climate change awareness, assessing vulnerability, and developing adaptation plans. In addition, this project made a major contribution to the Climate Change Adaptation Library (Halofsky and Peterson 2016, 2017; http://adaptationpartners.org/library.php), which is being used by land managers throughout the western United States.

Organizational capacity to address climate change requires building the institutional knowledge and ability of leadership, resource specialists, and management units to address climate change in agency operations. Training and education were built into the NRAP process through workshops and webinars that provided information about the effects of climate change on water resources, fisheries, vegetation, disturbance, wildlife, recreation, ecosystem services, and cultural resources. The workshops introduced climate tools and processes for assessing vulnerability and planning for adaptation. The participation of over 250 people in climate change workshops ensured the direct participation of agency employees in the NRAP, including their contributions to the assessment and adaptation options.

The NRAP science-management partnership and process were as important as the products that were developed, because partnerships are the cornerstone of successful agency responses to climate change. Land management agencies in the western United States have embraced partnerships in order accomplish their mission of sustainable resource stewardship, because diverse perspectives, timely feedback, and consensus building improve the likelihood of successful implementation of plans and projects, and reduce the likelihood of appeals and litigation. *Working across boundaries*—both geographic and sociopolitical—ensures that consistent approaches are applied to natural resource issues, especially for those resources that overlap jurisdictions (water, large animals, fish, etc.) (Olliff and Hansen 2016). Building enduring relationships and developing a shared vision are becoming more common in federal resource management and beyond. The NRAP process allowed the U.S. Forest Service (USFS) and National Park Service to achieve unit-level compliance in their agency-specific climate responses, but the influence of the project on broad landscapes inside and outside their borders was a more important outcome. The *all-lands approach* used in the Northern Rockies is critical for improving the resilience of ecosystems and organizations in the future.

The science-management dialogue created by NRAP identified management practices that are useful for increasing resilience and reducing stressors from climate change and other causes. Although implementation of all options developed in the NRAP process may not be feasible, resource managers can draw from the menu of options as needed. Some adaptation strategies and tactics can be implemented on the ground now, whereas others may require changes in policies and practices, or can be implemented when management plans are revised or as threats become more apparent.

Various components of the NRAP process identified *information gaps and uncertainties* important to understanding climate change vulnerabilities and adaptation. This is especially relevant for developing monitoring and research that will decrease uncertainties inherent to management decisions. In addition, current monitoring programs that provide information for detecting climate change effects—as well as indicators, species, and ecosystems that require additional monitoring—were identified for some components of the assessment. Working across multiple jurisdictions and boundaries will allow NRAP participants to increase collaborative monitoring and research on climate change effects and on the effectiveness of adaptation.

12.1.2 Implementation: The Path Forward

The NRAP built on previous science-management partnerships by creating an inclusive forum for local and regional stakeholders to address issues related to climate change vulnerability and adaptation. Although this partnership was conducted at the regional scale, more work is needed to truly achieve an all-lands approach to adaptation. Agencies used this process to share different approaches and experiences, and opportunities for creating a collaborative adaptation plan were explored. In the future, it would be valuable to *develop partnerships around specific resource issues and implement adaptation options accordingly*. In addition, *working at the subregional scale* would address specific management issues at a more geographically appropriate scale. Finally, because the NRAP process was able to engage only a subset of the total federal workforce represented by federal agencies, *continued internal and external communication* focused on climate-smart thinking, planning, and management will be necessary to ensure consistency and compliance with agency mandates.

Implementing adaptation options is challenging, although it builds on a solid foundation of planning principles and management practices that are already climate smart. Thinning of dry forests to increase vigor and reduce fuel loadings, restoration of riparian areas to enhance cold-water fish habitat, and removing roads from floodplains and other vulnerable locations are climate-smart practices that have been part of sustainable resource management for many years. Broader implementation will occur gradually over time in response to new policies, plan revisions, and programmatic directives. In some cases, extreme weather events (e.g., prolonged droughts) and major ecological disturbances (e.g., large wildfires) may provide the motivation to implement climate-smart actions. As previously stated, collaboration among landowners and management agencies will produce more successful adaptation outcomes than operating independently.

Landscape management strategies provide a context for decision making in which managers can be transparent in decisions to apply any given strategy or tactic. Appropriate adaptation options must consider resource conditions, social and ecological values, and likelihood of successful outcomes in a warmer climate (Peterson et al. 2011). The use of planning teams to develop resource-specific critical questions, and their responses to those questions, can inform broadly applicable management strategies (Figs. 12.1 and 12.2). The Climate Project Screening Tool (Morelli et al. 2012) and similar straightforward approaches can be applied to both strategic and project management.

Adaptation options that provide benefits to multiple resources will often have the greatest benefit in a particular landscape (Halofsky et al. 2011; Peterson et al. 2011; Raymond et al. 2014; Halofsky and Peterson 2016, 2017). However, some adaptation options involve tradeoffs and uncertainties that need further exploration. Assembling an interdisciplinary team to tackle this issue will be critical for assessing risks and developing risk management options.

The climate change vulnerability assessment and adaptation approach developed by the NRAP can be used by the USFS, National Park Service, and other organizations in many ways (Table 12.1):

- *Landscape management assessments/planning*: The vulnerability assessment provides information on departure from desired conditions and best science on effects of climate change on resources for inclusion in planning assessments.

12 Toward Climate-Smart Resource Management in the Northern Rockies

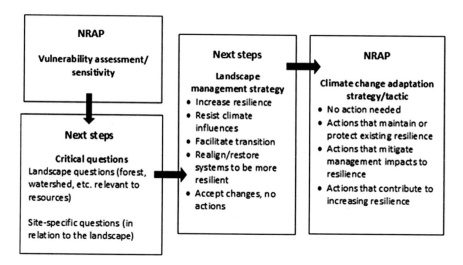

Fig. 12.1 General framework for using NRAP vulnerability assessment and adaptation information to ask critical questions and develop a landscape management strategy

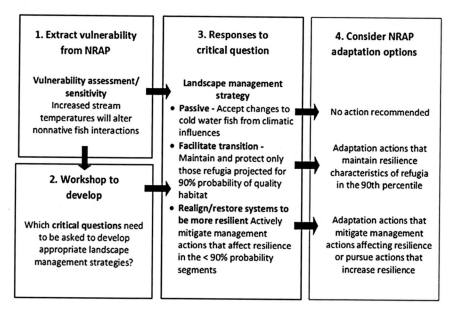

Fig. 12.2 Example of how to answer critical questions and develop a landscape management strategy for cold-water fish

Table 12.1 Example of how information on climate change vulnerability and adaptation can be used in land management applications in dry forests

Vulnerability and adaptation information	Land management application
Sensitivity to climatic variability and change	
Potential conversion to grassland	Forest planning: assessment phase
Many ponderosa pine forests have converted to Douglas-fir types because of fire exclusion and are therefore more susceptible to future fires	National Environmental Policy Act (NEPA) project analysis: existing condition and best science on effects of climate change on resource
Adaption strategy	
Restore fire-adapted ponderosa pine stand conditions in order to facilitate transition	Forest planning: desired conditions
	Project NEPA analysis: purpose and needs
Adaptation tactics	
Reduce competition from Douglas-fir and grand fir (thin, burn) in current mature pine stands	Forest planning: objectives
Conduce frequent understory burning	Project NEPA analysis: project design features and other mitigation
Retain current mature and older ponderosa pine stands	
Plant ponderosa pine where it has been lost	

Adaptation strategies and tactics provide desired conditions, objectives, standards, and guidelines for land management plans and general management assessments.

- *Resource management strategies:* Vulnerability assessment and adaptation strategies and tactics can be used to incorporate best science into conservation strategies, fire management plans, infrastructure planning, and State Wildlife Action Plans.
- *National Environmental Policy Act (NEPA) analysis for projects:* The vulnerability assessment provides best available science for documentation of resource conditions, effects analysis, and development of alternatives. Adaptation strategies and tactics provide mitigation and design tactics at specific locations.
- *Monitoring plans:* The vulnerability assessment identifies knowledge gaps that can be addressed by monitoring in broad-scale strategies, plan-level programs, and project-level data collection.

We are optimistic that climate change awareness, climate-smart planning and management, and implementation of adaptation in the Northern Rockies will progress quickly in the years ahead. We anticipate that:

- The effects of climate change on natural and human systems will be continually assessed.
- Monitoring activities will include indicators to detect the effects of climate change on species and ecosystems.

- Agency planning processes will provide opportunities to manage across boundaries.
- Managers will implement climate-informed practices in long-term planning and management.
- Restoration activities will be implemented in the context of a changing climate.
- Institutional capacity to manage for climate change will increase within federal agencies and other organizations.

Some climate-related changes in natural resources will be gradual, and others will occur abruptly. Timely actions will be necessary in some cases to protect critical ecosystem components and valued species, especially in the face of increasing frequency of extreme events (e.g., drought) and disturbances (e.g., wildfire, insect outbreaks). It will be critical for federal agencies and other organizations to share their experiences with implementation of climate-smart planning and management, building on successes and modifying approaches as necessary.

Federal agencies have demonstrated leadership in the implementation of ecosystem-based management, ecological restoration, and conservation of biological diversity over the past 30 years. Incorporating these paradigms in sustainable resource management required an enormous shift in organizational focus, whereas incorporating climate change will require mostly "fine tuning" of existing programs and practice. We are confident that resource managers and scientists in the Northern Rockies will improve the resilience of both ecosystems and organizations to a warmer climate, thus ensuring long-term sustainability.

References

Halofsky, J. E., & Peterson, D. L. (2016). Climate change vulnerabilities and adaptation options for forest vegetation management in the northwestern USA. *Atmosphere, 7*(3), 46. doi:10.3390/atmos7030046.

Halofsky, J. E., & Peterson, D. L. (Eds.). (2017). Vulnerability and adaptation to climate change in the Blue Mountains (General Technical Report PNW-GTR-939). Portland: U.S. Forest Service, Pacific Northwest Research Station.

Halofsky, J. E., Peterson, D. L., O'Halloran, K. A., & Hoffman, C. H. (Eds.). (2011). Adapting to climate change at Olympic National Forest and Olympic National Park (General Technical Report PNW-GTR-844). Portland: U.S. Forest Service, Pacific Northwest Research Station.

Morelli, T. L., Yeh, S., Smith, N., et al. (2012). Climate Project Screening Tool: An aid for climate adaptation (Research Paper PSW-RP-263). Albany: U.S. Forest Service, Pacific Southwest Research Station.

National Park Service (NPS). (2010). *National Park Service climate change response strategy.* http://www.nps.gov/orgs/ccrp/upload/NPS_CCRS.pdf. 15 Dec 2016.

Olliff, S. T., & Hansen, A. J. (2016). Challenges and approaches for integrating climate science into federal land management. In A. J. Hansen, W. B. Monahan, D. M. Theobald, & S. T. Olliff (Eds.), *Climate change in wildlands* (pp. 33–52). Washington, DC: Island Press.

Peterson, D. L., Millar, C. I., Joyce, L. A., et al. (2011). *Responding to climate change in national forests: A guidebook for developing adaptation options* (General Technical Report PNW-GTR-855). Portland: U.S. Forest Service, Pacific Northwest Research Station.

Raymond, C. L., Peterson, D. L., & Rochefort, R. M. (Eds.). (2014). Climate change vulnerability and adaptation in the North Cascades region (General Technical Report PNW-GTR-892). Portland: U.S. Forest Service, Pacific Northwest Research Station.

Swanston, C., Janowiak, M., Iverson, L., et al. (2011). *Ecosystem vulnerability assessment and synthesis: A report from the Climate Change Response Framework project in northern Wisconsin* (General Technical Report NRS-82). Newtown Square: U.S. Forest Service, Northern Research Station.

U.S. Forest Service (USFS). (2010). *National roadmap for responding to climate change.* http://www.fs.fed.us/climatechange/pdf/roadmap.pdf. 15 Dec 2016.

U.S. Forest Service (USFS). (2012). *A performance scorecard for implementing the Forest Service climate change strategy.* http://www.fs.fed.us/climatechange/pdf/performance_scorecard_final.pdf. 15 Dec 2016.

Index

A
A1B scenario, 39
Abies grandis, 7, 66, 70
Abies lasiocarpa, 7, 73, 74, 129
Absaroka range, 28
Acer glabrum, 105
Achnatherum nelsonii, 104
Achnatherum richardsonii, 104
Adaptation, 2–6, 33
Adaptive capacity, 4, 5, 106, 110, 148, 179, 191–194, 204
Adaptive management, 54, 85, 88
Agriculture, 13, 19, 109, 120, 195, 197, 202, 205, 206
Air quality, 178, 192, 200, 201, 205
Alces alces, 8, 152
Alnus viridus subsp. *sinuata*, 105
Alpine larch, 7, 9, 65, 73, 75, 76, 87
Amelanchier alnifolia, 8, 105
American beaver, 34, 52, 146, 149, 150, 160
American pika, 150
Amphibians, 13, 38, 145, 150
Anaconda-Pintler range, 75
Anaxyrus boreas, 160
Andropogon gerardii, 110
Andropogon spp., 109
Annosus root disease, 70, 131
Annual grasses, 101, 102, 106–108, 135
Antilocapra americana, 153
Aquatic habitat, 33, 49, 54, 181, 196, 197, 205
Aquatic macroinvertebrates, 38
Aquifers, 33, 34
Arceuthobium, 131
Area burned, 64, 72, 79, 81, 118, 120–122, 128, 212

Armillaria ostoyae, 131
Armillaria root rot, 70
Armillaria spp., 70
Artemisia, 100, 147, 200
Artemisia arbuscula, 100
Artemisia cana, 100
Artemisia nova, 100
Artemisia tridentata, 134, 153
Artemisia tridentata ssp. *tridentata*, 100
Artemisia tridentata ssp. *vaseyana*, 11, 99, 100, 153
Artemisia tridentata ssp. *wyomingensis*, 99, 100
Artemisia tripartita, 100
Artifacts, 210–214, 216, 217
Assisted migration, 54

B
Badlands, 12, 13
Bark beetles, 64, 116, 122–126
Base flows, 33
Basin big sagebrush, 100, 107, 108
Batrachochytrium dendrobatidis, 160
Beartooth Plateau, 44
Beartooth ranges, 28
Beaverhead-Deerlodge National Forest, 6, 7, 175
Behavioral plasticity, 144, 152
Big sagebrush, 11, 99–101, 106–109, 134, 153
Biogeochemical processes, 122
Biological diversity, 134
Bitterroot National Forest, 216
Bitterroot Range, 6, 75, 156
Bitterroot Valley, 8–10

© Springer International Publishing AG 2018
J.E. Halofsky, D.L. Peterson (eds.), *Climate Change and Rocky Mountain Ecosystems*, Advances in Global Change Research 63,
DOI 10.1007/978-3-319-56928-4

Black cottonwood, 7, 9, 77, 78
Black sagebrush, 100, 105, 106
Blackfoot River, 10, 47
Blue grama, 12, 110
Bluebunch wheatgrass, 104, 107, 109, 134, 135
Bluestem, 12, 109, 110
Bluetongue, 153
Boating, 172, 182, 184, 194
Bonasa umbellus, 147, 159
Bonneville Basin, 41
Boundary layer, 144
Bouteloua dactyloides, 110
Bouteloua gracili, 110
Bouteloua spp., 109
Brachylagus idahoensis, 153, 154
Brewer's sparrow, 147, 156, 157, 161
Bridger-Teton National Forest, 7, 198
Bromus arvensis, 107
Bromus japonicus, 101
Bromus tectorum, 101, 120, 158, 200
Brook trout, 8, 11, 30, 38, 40–42, 45–48, 51
Brown trout, 38, 40, 51
B2 scenario, 65, 103
Buffalograss, 110
Bull trout, 8, 9, 30, 38, 40–42, 44, 46–48, 50, 51, 196
Bunchgrass, 9, 12, 100
Burn severity, 135

C

Cabinet Ranges, 75
Camping, 170, 172–174, 178
Canada lynx, 9, 11, 145, 146, 150, 162
Canker disease, 131
Carbon cycling, 122
Carbon dioxide (CO_2), 63, 101, 102, 104, 107–109, 134, 200
Carbon sequestration, 84, 190, 192, 202–204, 206
Carbon stocks, 202, 203
Carbon storage, 84, 201–204
Caribou-Targhee National Forest, 7, 198
Castor canadensis, 34, 52, 149, 150
Cavity nester, 157, 158
Ceanothus velutintis var. *velutinus*, 105
Centaurea melitensis, 134
Center of runoff timing, 29–30
Centrocereus urophasianus, 8, 98, 147, 157, 158
Cervus canadensis, 154

Cheatgrass, 98, 99, 101, 104, 107–109, 120, 135, 158, 200
Chokecherry, 12, 105
Choristoneura occidentalis, 68
Chrosomus eos, 49
Chrysothamnus viscidiflorus, 105
Chytrid fungus, 160
Clark's nutcracker, 73, 76
Climate Shield, 38, 51
Cold-water fish, 38–42, 44–47, 50–54, 149, 172, 181, 224, 225
Cold-water habitats (CWHs), 42, 44–46
Colorado River, 11, 156
Columbia River, 7, 11, 150
Columbia spotted frog, 146, 159, 160
Connectivity, 51, 54, 150, 151, 156, 161, 162, 214
Continental Divide, 80, 131, 214
Cool-season (C3) grasses, 104, 109, 110
Corynorhinus townsendii, 154
Coupled Model Intercomparison Project (CMIP), 18–20, 22, 30, 39
Cronartium ribicola, 8, 116, 127, 214
Crown fires, 34, 69, 71, 80, 119, 121, 128
Cultural resources, 4, 210–217, 222
Currant, 10, 105
Custer-Gallatin National Forest, 49, 214
Cutthroat trout, 8, 9, 11, 12, 30, 38, 41, 42, 44–47, 50, 51, 180, 181, 196

D

Dakota Prairie National Grassland, 6, 49
Dams, 30, 34, 202
Dasiphora fruticosa, 134
Dendroctonus brevicomis, 124
Dendroctonus ponderosae, 8, 64, 122, 214
Dendroctonus pseudotsugae, 68, 123
Dendroctonus rufipennis, 74, 123
Desert, 98, 99, 153
Diseases, 111, 116, 130–133, 147, 152, 153, 161, 162, 198, 204
Disturbance regimes, 60, 62, 63, 76, 81, 107, 116, 117, 147, 177, 200
Dothistroma needle blight, 133
Dothistroma septosporum, 133
Douglas-fir, 7–11, 65, 66, 68–71, 73, 79, 80, 83, 86, 88, 123, 129–131, 147, 161, 226
Douglas-fir beetle, 68
Douglas-fir tussock moth, 68
Downscaling, 5, 19, 125

Index

Drought, 11, 22, 26, 27, 33, 34, 49, 62–64, 66, 68–74, 76–80, 84, 86–88, 102, 104–110, 120, 122, 123, 128, 131, 132, 135, 150, 157, 159, 181, 182, 195, 197, 202, 211–213, 216, 224, 227
Drought stress, 75, 122, 134, 198
Drought tolerance, 69, 75
Dwarf mistletoes, 130–133

E

Ecologically based invasive plant management (EBIPM), 110, 111
Ecosystem models, 62, 84
Ecosystem services, 82, 83, 98, 130, 190–192, 194–206
Ectotherms, 38, 145
El Niño Southern Oscillation (ENSO), 120
Elaeagnus angustifolia, 134
Emissions scenarios, 19, 125
Endotherm, 144, 145, 147
Energy development, 98, 109
Engelmann spruce, 7, 9–11, 65, 73–76, 80, 83, 87, 129, 130
Environmental DNA (eDNA), 54
Ericameria nauseosa, 105
Erosion, 34, 53, 104, 106, 111, 177, 181, 182, 192, 198, 201, 202, 205, 210–213
Euphorbia esula, 101
Evapotranspiration, 22, 26, 102
Exposure, 4, 5, 148, 182, 199, 211
Extreme weather, 77, 121, 127, 216, 224

F

Festuca campestris, 104
Festuca idahoensis, 10, 104
Field brome, 107
Fir engraver, 123
Fire behavior, 121, 122, 133
Fire exclusion, 8, 10, 63, 64, 68–71, 75, 77, 79, 81, 86, 87, 101, 104, 105, 108, 111, 120, 121, 133, 161, 226
Fire frequency, 8, 77, 78, 105–109, 118, 119, 121, 128, 129, 205, 217
Fire intensity, 117, 120, 122
Fire management, 88, 120, 178, 217, 226
Fire regime, historical, 119
Fire regimes, 76, 87, 100, 109–111, 118, 119, 121, 122, 155, 158, 200
Fire return intervals, 100, 101, 106, 107, 109
Fire season, 27, 70, 118, 120, 121, 198, 201
Fire severity, 68, 86, 118, 119, 121, 122, 202
Fire suppression, 67, 87, 120, 212, 217

Fire, high-intensity, 133
Fire, low-intensity, 69, 133, 178
Fire, moderate-intensity, 65, 71, 72, 86, 174, 181
Fisher, 146, 148, 151, 152, 161
Fisheries, 4, 52–54, 84, 195, 222
Fishing, 11, 53, 54, 172–174, 180, 182, 183, 196, 197
Flammulated owls, 146, 157, 161
Flathead National Forest, 6, 198, 203
Flathead River, 47
Flavivirus spp., 147
Flooding, 32, 33, 78, 132, 185, 197, 202, 211–214, 217
Floodplains, 33, 52, 53, 78, 224
Forage, 98, 107, 111, 134, 151, 154, 155, 160, 181, 192, 199, 200, 205
Forest productivity, 120
Forest products, 173, 174, 181–183, 197
Fragmentation, 64, 107, 161, 177, 200, 205
Fraxinus pennsylvanica, 12, 78, 79
Frosts, 62, 63, 66, 70, 71, 104
Fuel reduction, 4, 111
Fuel treatments, 53, 71, 85, 120, 159, 198, 201, 202, 206
Fuels, 53, 63, 64, 68–71, 75, 79, 81, 84, 86, 101, 104, 106, 109, 117–122, 128, 133, 159, 191, 192, 216, 218
Fuels management, 120, 198, 202
Fungal pathogens, 128, 204, 214
Fungi, 74, 122, 124, 131–133, 215
Future range of variability, 81

G

Genetic diversity, 63, 64, 87, 88, 192
Germination, 69, 73, 74, 78, 107, 127, 132, 133
Glacier National Park, 6, 9, 10, 28, 29, 72, 158, 170
Glaciers, 2, 4, 8, 9, 26, 28, 29, 172, 178
Global climate models (GCMs), 18, 19, 27, 38, 40, 82, 103
Grama, 109, 110
Grand fir, 7–9, 65, 66, 68, 70, 71, 73, 79, 83, 86, 131, 226
Grand Teton National Parks, 4
Grasslands, 2, 6, 9–13, 18, 20, 22, 47, 49, 52, 53, 65, 73, 81, 87, 98, 100–102, 104, 107, 109–111, 120, 121, 135, 146, 154, 161, 196, 200, 203, 226
Grazing, 2, 34, 52–54, 73, 87, 100, 101, 104, 106–108, 111, 120, 161, 193, 196, 200, 205, 212

Great Plains, 78, 100–102, 109, 110, 158, 210
Greater sage-grouse, 8, 11, 98, 100, 146, 147, 157, 158, 161
Greater Yellowstone Area (GYA), 6, 7, 11, 18, 20, 22, 65, 76, 83, 101, 103, 106–108, 121, 128, 146, 155, 156, 161, 170, 198
Green ash, 12, 65, 78, 87
Green rabbitbrush, 105
Green River, 11
Greenhouse gases, 18, 19, 125, 202
Grinnell Glacier, 28, 29
Groundwater, 29, 44, 49, 53, 196
Growing season, 62, 63, 69–71, 78–80, 102, 127, 195
Growth, 11, 38, 50, 60–63, 68–74, 78, 82, 102–106, 108, 111, 118, 123–126, 131–133, 146, 152, 155, 177, 195, 199, 200, 202, 203
Gulo gulo, 11, 155, 156

H
Habitat connectivity, 51
Harlequin ducks, 146, 158
Headwater streams, 49, 196
Helena National Forest, 6, 198, 213, 214, 217
Herbivory, 13, 77, 104, 107, 108
Hesperostipa comata, 10, 104
Heteribasidion annosum, 70
Hibernation, 154
Hiking, 170, 172–174, 178
Historic building, 214, 216, 218
Historical range and variability (HRV), 81, 82
Historical sites, 216
Histrionicus histrionicus, 158
Holidiscus discolor, 8, 105
Huckleberries, 8, 181, 216
Hunting, 13, 170, 173, 174, 180, 181, 183, 184, 210, 213
Hydroelectric power, 194
Hydrology, 2, 4, 19, 20, 26–28, 30, 32, 33, 38, 39, 41, 52, 53, 60, 78, 82, 120, 195, 201, 213, 216, 217

I
Idaho, 5–9, 28, 30, 41, 44, 47, 51, 77, 79, 104, 105, 120, 148, 152, 153, 156, 173, 176, 196, 197, 201, 203, 204, 216
Idaho fescue, 10, 11, 104
Idaho Panhandle National Forest, 198, 203, 204
Implementation, 5, 6, 18, 53, 85, 88, 223–227
Infrastructure, 20, 22, 30, 33, 109, 170, 185, 192, 195, 197–199, 201, 205, 226

Insect outbreaks, 2, 11, 63, 174, 203, 205, 206, 227
Insects, 2, 11, 12, 18, 64, 68–71, 74, 76–79, 83, 88, 107, 111, 116, 125, 133, 154, 157, 158, 195, 197, 215, 218
Intergovernmental Panel on Climate Change (IPCC), 4, 18, 26, 32, 85, 213
Invasive plants, 107, 110, 134, 200, 205
Invasive species (invasives), 84, 85, 100, 101, 107, 109, 110, 135, 211
Ips spp., 123–125

J
Japanese brome, 101, 109

K
Kentucky bluegrass, 109
Koeleria macrantha, 104
Kootenai National Forest, 6, 198
Kootenai River, 8, 51, 194

L
Lake trout, 12, 40, 41
Laminated root rot, 75
Landscape heterogeneity, 80, 81, 86, 88
Landslides, 198, 201, 202
Larix lyallii, 7, 9, 75, 76
Larix occidentalis, 7, 66, 68, 69
Leafy spurge, 101, 109
Lepus americanus, 145, 150
Lewis and Clark National Forest, 6, 198
Life history traits, 98, 148
Lightning, 117, 118, 120
Limber pine, 7, 8, 10, 65, 73, 87, 127
Little bluestem, 12, 110
Little Missouri River, 12
Livestock, 34, 52, 53, 104–109, 111, 192, 194, 199, 200, 205, 212
Lodgepole pine, 7, 8, 10–12, 65, 66, 71–73, 75, 76, 80, 83, 86, 88, 89, 123, 128–131, 133, 155
Lolo National Forest, 6, 198
Low sagebrushes, 100, 105, 106
Lynx canadensis, 9, 150, 151

M
Malacosoma spp., 78
MC2 model, 64, 66
Medusahead, 135
Migration, 38, 40, 50, 53, 54, 63, 81, 150
Minerals, 69, 71, 199, 205

Index

Mining, 2, 191, 192, 194, 199, 210, 212, 214
Mississippi River, 11
Missoula Valleys, 8–10
Missouri River, 10–13, 41, 156, 194, 195
Mixed grass prairie, 109, 110
Monitoring, 111, 148, 162, 172, 217, 218, 223, 226
Montana, 5, 6, 8–12, 27, 30, 41, 44, 51, 75, 77, 79, 83, 86, 105, 107–109, 120, 134, 148, 151, 152, 156, 158, 160, 193–197, 199, 213–216
Montane grasslands, 100–102, 104
Moose, 146, 152, 162
Mountain ash, 105
Mountain big sagebrush, 99, 106, 108, 109, 153, 154
Mountain hemlock, 7, 65, 73, 75, 76, 87
Mountain pine beetles (MPB), 8, 10–12, 64, 68, 72, 73, 76, 77, 80, 81, 83, 88, 117, 122, 124, 126, 128–130, 214, 215
Mountain quail, 146, 158, 161
Municipal water supply, 34, 194–196

N

Narrowleaf cottonwood, 77
National forests, 2–6, 33, 39, 42, 44, 49, 51, 131, 148, 170, 172, 174–176, 178, 197, 198, 202–204, 212, 213, 216, 217, 222
National Park Service (NPS), 6, 11, 12, 149, 155, 170, 210, 214, 222–224
National parks, 2–4, 6, 39, 51, 170, 171, 178, 179, 216, 217, 222
National Visitor Use Monitoring (NVUM), 172
Native Americans, 2, 13, 192, 196, 210, 213, 214, 216
Needle-and-thread, 10, 12, 104, 107, 109
Net primary productivity (NPP), 102, 103
Nez Perce-Clearwater National Forest, 6, 198
Nonnative species, 38, 41, 53, 54, 60, 104–106, 108, 109, 120, 134, 135, 155, 161, 196, 199
North American Monsoon System, 27
North Dakota, 5, 12, 109, 195, 196, 199, 201
North Fork Blackfoot River, 47
Northern bog lemming, 146, 153
Northern redbelly dace, 49
NorWeST, 39–41, 43
Nutrient cycling, 120, 190, 201

O

Oceanspray, 8, 105
Ochotona princeps, 150

Odocoileus virginianus, 12, 152, 154
Odocoleus hemionus hemionus, 154
Oil and gas development, 107, 194, 199
Oncorhynchus clarkii, 8, 38, 180, 196
Oncorhynchus clarkii bouvieri, 12
Oncorhynchus clarkii lewisi, 9, 41
Oncorhynchus mykiss, 8, 38, 181
Oreortyx pictus, 158
Orgyia pseudotsugata, 68
Otus flammeolus, 157

P

Pacific Decadal Oscillation (PDO), 120
Pacific Northwest, 4, 19, 27, 38, 121
Palouse prairie, 104
Parasite, 152
Pascopyrum smithii, 104
Pathogens, 69, 71, 74, 77, 107, 123, 128, 131–133, 204, 214
Peakflow, 30, 33, 181
Pekania pennanti, 148, 151, 152
Pelage change, 145
Perennial grasses, 104, 106, 107, 135
Pesticide, 154
Phaeocryptopus gaeumannii, 132
Phellius weirii, 75
Phenology, 62, 117, 146, 148, 162
Phenotypic plasticity, 64
Photosynthesis, 63, 190
Picea engelmannii, 7, 74, 75
Pinus albicaulis, 76, 77, 127, 214
Pinus contorta var. *latifolia*, 66, 72, 73
Pinus flexilis, 7, 8, 73, 127
Pinus monticola, 7, 68–70, 127
Pinus ponderosa, 7, 63, 66–68, 102, 119
Plains cottonwood, 77, 78
Planning, 2, 4, 6, 18, 19, 82, 84, 85, 170, 177, 194, 200, 218, 222, 224, 226, 227
Poa pratensis, 109
Poa secunda, 104
Pollinators, 111
Ponderosa pine, 7, 9, 11, 12, 63, 66, 68, 70, 71, 79, 80, 83, 86, 88, 102, 119, 129, 130, 158, 161, 226
Populus angustifolia, 77
Populus deltoides, 12, 77
Populus tremuloides, 7, 77, 158
Populus trichocarpa, 77
Prairie, 12, 13, 47, 49, 109, 153, 161, 196
Prairie junegrass, 104
Prescribed fire, 87, 88, 110, 161
Productivity, 8, 9, 50, 62, 70, 71, 73, 75, 80, 83, 102, 103, 105, 106, 108–110, 120, 131, 133, 150, 161, 181, 200, 205

Pronghorn, 13, 146, 153
Prunus virginiana, 12, 105
Pseudogymnoascus destructans, 162
Pseudoroegneria spicata, 104, 134
Pseudotsuga menziesii, 7, 66, 68, 147
Pygmy nuthatch, 146, 158, 159
Pygmy rabbit, 146, 153, 154, 161, 162

Q
Quaking aspen, 7, 9, 11, 68, 77, 80, 87, 146, 158, 161

R
Radiative forcing, 19
Rainbow trout, 8, 11, 30, 38, 41, 42, 51, 181
Rain-on-snow events, 26, 30
Rana luteiventris, 159, 160
Rangelands, 18, 98–102, 104, 106, 110, 111, 199–202, 205
Recreation, 4, 10, 30, 134, 169–185, 190–196, 210, 212, 215, 222
Recreation Opportunity Spectrum (ROS), 170, 185
Red belt, 131
Red River, 12, 109
Red River basin, 109
Refugia, 22, 42, 45–47, 50, 51, 53, 54, 81, 88, 181
Regeneration, 60, 62, 63, 68–72, 75–78, 80, 88, 98, 105, 107, 123, 127, 128, 202, 216
Representative concentration pathways (RCPs), 19–21, 39, 65, 125, 126
Reservoir, 10, 13, 22, 34, 125, 174, 182–184, 194–196, 202, 205
Resilience, 33, 52, 53, 69, 78, 79, 81, 88, 98, 99, 110, 111, 148, 160, 161, 223, 227
Resistance, 68, 73, 86, 87, 89, 98, 99, 104, 127, 133, 135
Restoration, 4, 5, 51–54, 70, 77, 79, 81, 83, 85–88, 149, 197, 202, 224, 227
Ribes spp., 105, 127, 128
Richardson's needlegrass, 104
Riparian, 4, 9, 33, 49, 51–54, 71, 77, 78, 85, 106, 134, 146, 150, 154, 160, 197, 202, 205, 224
Risk assessment, 33, 118, 124, 132, 133, 215, 216
Risk management, 85, 224
Roads, 30, 33, 34, 52–54, 89, 109, 177, 196, 201, 224

Rocky Mountain elk, 154
Rocky Mountain Front, 45
Rocky Mountain maple, 105
Rocky Mountain mule deer, 154
Root disease, 70, 83, 86, 130, 131, 133
Rough fescue, 102, 104
Rubber rabbitbrush, 105
Rubus parviflorus, 105
Ruffed grouse, 146, 147, 159, 161
Runoff, 29, 195, 196, 211
Russian olive, 134

S
Sagebrush, 10, 12, 99–102, 104–109, 120, 122, 147, 153, 156, 157, 161, 200
Salix scouleriana, 105
Salmo trutta, 38
Salmon River, 7
Salmonids, 38, 47, 49, 51
Salvelinus confluentus, 8, 38, 196
Salvelinus fontinalis, 8, 38
Salvelinus namaycush, 12, 40
Sandberg bluegrass, 104
Schizachyrium scoparium, 12, 110
Science-management partnerships, 2, 4, 222, 223
Scolytus ventralis, 123
Scouler willow, 105
Sediment yield, 33
Seed dispersal, 63, 73, 75
Seedling establishment, 62, 78, 87
Seedlings, 66, 68, 69, 71–75, 78, 87, 89, 107
Sensitivity, 2, 4, 5, 33, 45, 52, 148–161, 197, 200, 203, 226
Serviceberry, 7, 12, 105
Shoshone National Forest, 2, 7, 198, 210
Shrubby cinquefoil, 134
Shrublands, 53, 81, 98, 105, 106, 108, 111, 118, 120, 153
Silver sagebrush, 100, 106, 107
Sitka alder, 105
Sitta pygmaea, 158, 159
Skiing, 170, 171, 173–175, 179, 180, 184
Smith River, 10
Smoke, 83, 128, 133, 174, 178, 183, 201
Snake River, 11, 158
Snow, 11, 26, 28–30, 60, 69, 74–76, 107, 125, 127, 144–146, 149–151, 153–156, 171, 173–175, 178–185, 202, 204, 216
Snow residence time, 28
Snow water equivalent (SWE), 28, 29
Snowberry, 99, 105

Index

Snowmobiling, 173, 174, 179, 180
Snowpack, 22, 26–29, 32–34, 38, 52, 63, 64, 73, 75, 76, 78, 104, 118, 120, 125, 146, 150, 152, 153, 155, 156, 162, 174, 175, 181–183, 213
Snowshoe hare, 145, 150, 151
Social vulnerability, 192
Social Vulnerability Index (SoVI), 192
Soil, 11, 13, 62, 66, 68, 69, 71–74, 76–79, 84, 86, 100–102, 104, 105, 108, 109, 111, 120, 125, 132–135, 146, 154, 190, 198, 201–205, 212, 213
Sorbus scopulina, 105
Souris River, 12
South Dakota, 6, 12
Special Report on Emissions Scenarios (SRES), 19
Species distribution models, 41, 61
Spineless horsebrush, 105
Spizella breweri, 147, 156, 157
Spotted knapweed, 134, 135
Sprouting, 63, 78, 100, 105–107, 128
Spruce beetle, 74, 75, 83, 123
Stream temperature, 38–45, 50–54, 172, 196
Streamflow, 18, 22, 26, 27, 29–33, 38–40, 49–52, 54, 78, 87, 149, 158, 160, 172, 180, 182, 183, 185, 192, 213
Subalpine fir, 7, 9, 11, 65, 73–76, 79, 80, 83, 87, 129–131
Subalpine forests, 76, 80, 83, 105, 119
Subnivean, 146, 153, 154
Succession, 8, 69–72, 74, 77, 80, 86, 87, 123, 131, 147, 149, 162
Summer low flows, 29, 32
Sustainability, 4, 98, 145, 200
Swiss needle cast, 132
Symphoricarpos albus, 105
Synaptomys borealis, 153
System for Assessing Vulnerability of Species (SAVS), 148

T

Taeniatherum caput-medusae, 135
Tallgrass prairie, 109, 110
Tamarisk, 134
Tamarix ramosissima, 134
Tent caterpillars, 78
Teton River, 4, 45
Tetradymia canescens, 105
Thermal cover, 154, 155
Thermodynamic regulation, 149
Thimbleberry, 105

Thinning, 81, 83, 86, 88, 123, 128, 131, 159, 198, 206, 224
Three-tip sagebrush, 100
Thuja plicata, 7, 68, 71
Timber, 2, 81, 83, 84, 120, 190, 191, 193, 197, 198, 202, 203, 205
Timber harvests, 83, 197
Tobacco brush, 105
Townsend's big-eared bat, 146, 154
Tree mortality, 62, 64, 83, 117, 122–129, 131
Tree regeneration, 63, 88
Treeline, 74–76, 86
Tsuga heterophylla, 7, 68, 71, 72
Tsuga mertensiana, 75

U

U.S. Forest Service (USFS), 4, 6, 7, 9, 13, 34, 39, 49, 51, 79, 83, 99, 100, 148, 170, 172, 174–177, 179, 198, 201–204, 210, 212, 213, 217, 223, 224
Uncertainty, 20, 26–28, 30, 32, 51, 54, 64, 75, 82, 83, 85, 88, 102, 148, 177, 182, 223, 224
Ungulate, 77, 98, 100, 104, 107, 108, 111, 146, 153–155

V

Vapor pressure deficit, 26
Variable infiltration capacity (VIC) model, 27, 28, 39
Vegetation mosaics, 154
Viewsheds, 200, 201, 205
Voltinism, 124–126
Vulnerability, 148
Vulnerability assessments, 2, 4–6, 146, 217, 218, 222, 224–226

W

Warm-season (C4) grass, 100
Water deficits, 77, 198
Water quality, 34, 51, 53, 84, 182, 192, 196, 197, 199, 201, 202, 205, 206
Water rights, 52
Water storage, 33, 34
Water supply, 26, 29, 52, 122, 162, 194, 195
Water-use efficiency, 63, 68
West Nile virus, 147, 158, 162
Western hemlock, 7, 8, 65, 68, 70–72, 83, 86, 133
Western larch, 7–10, 65, 66, 68, 69, 71, 73, 76, 79, 83, 86, 88, 133

Western needlegrass, 104
Western pine beetle, 124, 125
Western redcedar, 7–9, 65, 68, 70, 71, 73, 83, 86
Western Rockies, 6–8, 65, 86, 101, 105, 107, 108, 146, 161
Western spruce budworm, 11, 68, 74
Western toad, 146, 160
Western wheatgrass, 104, 110
Western white pine, 7–9, 65, 68–71, 73, 79, 83, 86, 88, 127
Westslope cutthroat trout, 9, 11, 12, 41
Wetlands, 34, 146, 160, 195, 197
White pine blister rust (WPBR), 8, 10, 69, 70, 73, 76, 80, 83, 86–88, 116, 127–130, 214
Whitebark pine, 7–11, 65, 73, 75–77, 80, 87, 88, 127–130, 214, 215
Whitefish Range, 75
Whitefish River, 47
White-nose syndrome, 162
White-tailed deer, 11–13, 146, 152, 154
Wild horses, 98

Wilderness, 2, 6, 8–10, 51, 53, 151, 177
Wildfire, 2, 6, 8, 10–13, 18, 26, 33, 39, 53, 64, 68, 69, 71–81, 83, 84, 86–88, 102, 108, 109, 116–122, 148, 154–159, 161, 172–174, 177, 178, 181–183, 192, 196–198, 201–205, 210–217, 224, 227
Wildland-urban interface, 10
Wildlife, 4, 11, 12, 81, 84, 88, 100, 107, 134, 144–147, 160–162, 172–174, 177, 180, 181, 183, 185, 191, 195, 215, 222, 226
Wind River Range, 28, 150
Windflows, 27
Wolverine, 11, 146, 155, 156, 162
Woodlands, 10, 67, 98, 158
Wyoming, 2, 6, 12, 41, 99, 100, 107–109, 135, 150, 160
Wyoming big sagebrush, 99, 100, 107, 108

Y
Yellowstone cutthroat trout, 12, 41
Yellowstone National Park, 11, 28, 45, 159, 178, 191, 216